T0191805

Kernel Ridge Regression in Clinical Research

Ton J. Cleophas • Aeilko H. Zwinderman

Kernel Ridge Regression in Clinical Research

 Springer

Ton J. Cleophas
Department Medicine
Albert Schweitzer Hospital
Sliedrecht, Zuid-Holland, The Netherlands

Aeilko H. Zwinderman
Department Biostatistics and Epidemiology
Academic Medical Center
Amsterdam, The Netherlands

ISBN 978-3-031-10719-1   ISBN 978-3-031-10717-7   (eBook)
https://doi.org/10.1007/978-3-031-10717-7

This Springer imprint is published by the registered company Springer Nature Switzerland AG
The registered company address is: Gewerbestrasse 11, 6330 Cham, Switzerland

Preface

IBM (international business machines) has published in its SPSS statistical software 2022 update a very important novel regression method entitled Kernel Ridge Regression (KRR). It is an extension of the currently available regression methods and is suitable for pattern recognition in high dimensional data, particularly when alternative methods fail. Its theoretical advantages are plenty and include the

- Kernel trick for reduced arithmetic complexity
- Estimation of uncertainty by Gaussians unlike histograms
- Corrected data-overfit by ridge regularization
- Availability of eight alternative kernel density models for data fit

A very exciting and wide array of preliminary KRR research has already been published by major disciplines (like studies in quantum mechanics and nuclear physics, studies of molecular affinity/dynamics, and atomization energy studies, but also forecasting economics studies, IoT (Internet of Things) studies for e-networks, plant stress response studies, and big data streaming studies). In contrast, it is virtually unused in clinical research. This edition is the first textbook and tutorial of kernel ridge regressions for medical and healthcare students as well as recollection/update bench and help desk for professionals. Each chapter can be studied as a standalone, and using real as well as hypothesized data, it tests the performance of the novel methodology against traditional regression analyses. Step-by-step analyses of over 20 data files stored at Supplementary Files at Springer Interlink are included for self-assessment. We should add that the authors are well qualified in their field. Professor Zwinderman is past president of the International Society of Biostatistics (2012–2015) and Professor Cleophas is past president of the American College of Angiology (2000–2002). From their expertise, they should be able to make adequate selections of modern KRR methods for the benefit of physicians, students and investigators. The authors have been working and publishing together for 24 years, and their research can be characterized as a continued effort to demonstrate that clinical data analysis is not mathematics but rather a discipline at the interface of biology and mathematics.

Sliedrecht, Zuid-Holland,   Ton J. Cleophas
The Netherlands
Amsterdam, The Netherlands   Aeilko H. Zwinderman

Contents

Chapter 1
Traditional Kernel Regression

Abstract With non-normal outcome data, that remain non-normal in spite of trans-formations (Likert scales is a notorious example), data distributions may be skewed, and nonparametric regression analysis may provide better data fit than traditional parametric models do. The AICs (Akaike Information Criterion goodness of fit tests) of the full loglikelihood model and the kernel model were respectively 1085 and 920, difference 165. This means, that the probability of the kernel regression model to minimize information loss is e $^{(165)/2}$ times less worse than the traditional loglikelihood model is.

Keywords Kernel regression · Non-normal data · Skewed data · Nonparametric regression · Akaike information criterion · Traditional loglikelihood regression model

1.1 Summary

With traditional regression methods, the outcome values are assumed to be normally distributed around the regression line/curve. With non-normal outcome data, that remain non-normal in spite of transformations (Likert scales is a notorious example), data distributions may be skewed, and nonparametric regression analysis may provide better data fit than traditional parametric models do. Methods including nonparametric regression are pretty new, and not yet widely applied. They include: kriging, otherwise called Gaussian process regression, decision trees, and bagged (bootstrap aggregated) regression trees, kernel regressions, and median regression, otherwise called robust regression. The AICs (Akaike Information Criterion goodness of fit tests) of the full loglikelihood model and the kernel model were respectively 1085 and 920, difference 165. This means, that the probability of the kernel regression model to minimize information loss is e $^{(165)/2}$ times less worse than the

---

Supplementary Information The online version contains supplementary material available at [https://doi.org/10.1007/978-3-031-10717-7_1].

full loglikelihood model is. Obviously, the kernel model performed endlessly better. Kernel regression provided a stronger goodness of fit than did traditional loglikelihood testing. However, no p-values are obtained. Traditional regression analysis produced two strong predictors of body surface, namely weight and height, and a borderline significant age effect of $p = 0.040$.

1.2   Introduction

With traditional regression methods, the outcome values are assumed to be normally distributed around the regression line/curve. With non-normal outcome data, that remain non-normal in spite of transformations (Likert scales is a notorious example), data distributions may be skewed, and nonparametric regression analysis may provide better data fit than traditional parametric models do. Methods including nonparametric regression are pretty new, and not yet widely applied. They include: kriging, otherwise called Gaussian process regression, decision trees, and bagged (bootstrap aggregated) regression trees, kernel regressions, and median regression, otherwise called robust regression. In this chapter kernel regression will be tested against traditional regression analysis. As a data example, body heights and weights were assessed as possible predictors of measured body surfaces in 90 persons.

1.3   Kernel Regression

An example will be given of data appropriate for kernel regression. Kernel regression is equivalent to radial basis function analysis. Kernel frequency distributions consist of multiple identical Gaussian curves, rather than histograms consistent of bins with different lengths. Kernel regression measures the relationship between x and y data, where the expectation of y is conditional not on all x-values but on locally weighted averages of subsets of consecutive x-values with a predefined bandwidth. The subsets are described by their means and standard errors. And, together, these means tend to produce a simple linear regression model. Obviously, any linear regression can be replaced with a kernel regression, but the method is particularly appropriate in case of seemingly nonlinear patterns, that, in the end, are linear after all. In many cases a (much) better fit for the data is provided by kernel regression than by linear regression.

probability distribution

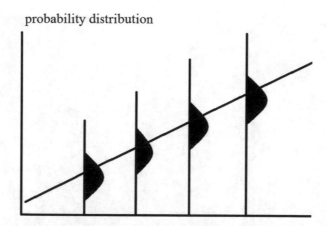

The above graph shows a linear regression model with the x-values sampled without error, and the y-values having uncertainties in the form of identical normal curves (Gaussian curves drawn in black). This linear regression model is parametric, because normal curves are symmetric around their mean.

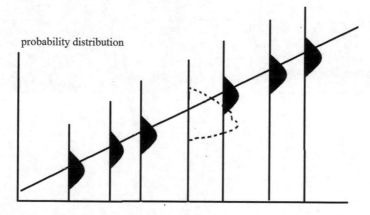

The above graph shows, how kernel regression works. The uncertainty of the y-values is not expressed in the form of Gaussian curves, but rather the add-up sum of multiple Gaussian curves. The area under curve of the dotted curve is the add-up sum of the six Gaussian areas under the curve (in black). It is called, a kernel density distribution. The corresponding x-values are subsets of all x-values of a data file. The subset given is within one bandwidth. Linear modeling of multiple bandwidths results in a linear model, even if the traditional linear model has a pretty poor linear fit.

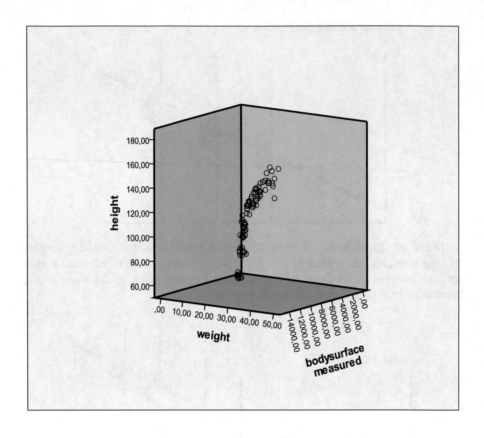

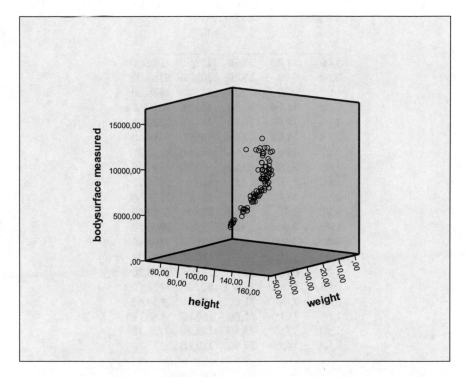

In the above graphs body heights and weights were assessed as possible pre-dictors of the measured body surfaces in 90 persons. The graphs show in the three dimensional scatter plots, that the relationships were, obviously, nonlinear. Body height and weight are generally assumed not to be linear with time but rather s-shape or biphasic. Kernel regression may be particularly advantageous for modeling such data. For analysis in SPSS statistical software we will use, from the SPSS module neural networks, radial basis neural (RBF) networks. The body surfaces of the 90 persons were calculated using direct photometric measurements. The data file is entitled "measured surface", and is in extras.springer.com. It is previously used by the authors in Machine learning in medicine a complete overview, Chap. 63, Springer Heidelberg Germany, 2015.

Variable

1	2	3	4	5
1,00	13,00	30,50	138,50	10072,90
0,00	5,00	15,00	101,00	6189,00
0,00	0,00	2,50	51,50	1906,20
1,00	11,00	30,00	141,00	10290,60
1,00	15,00	40,50	154,00	13221,60
0,00	11,00	27,00	136,00	9654,50
0,00	5,00	15,00	106,00	6768,20
1,00	5,00	15,00	103,00	6194,10
1,00	3,00	13,50	96,00	5830,20
0,00	13,00	36,00	150,00	11759,00
0,00	3,00	12,00	92,00	5299,40
1,00	0,00	2,50	51,00	2094,50
0,00	7,00	19,00	121,00	7490,80
1,00	13,00	28,00	130,50	9521,70
1,00	0,00	3,00	54,00	2446,20
0,00	0,00	3,00	51,00	1632,50
0,00	7,00	21,00	123,00	7958,80
1,00	11,00	31,00	139,00	10580,80
1,00	7,00	24,50	122,50	8756,10
1,00	11,00	26,00	133,00	9573,00
.	.	.	.	.
.	.	.	.	
.	.	.		

Var 1 = gender
Var 2 = age
Var 3 = weight (kg)
Var 4 = height (m)
Var 5 = body surface measured (cm^2)

With regression analyses the best fit regression equation is often used for making predictions from your data about future data. A problem with this procedure is, that the best fit regression equation is used for generating inter- and extrapolated data, and the fit is not always, what it should be, particularly with data, that tend to be nonlinear, like in the above graphs. It is accepted that some kind of goodness of fit test prior to prediction is recommendable, and, that, with multidimensional data, it is even possible to improve the predictive regression equation with the help of training samples.

First, the performance of kernel regression, as compared to traditional log likelihood modeling will be assessed. Akaike's information criterion (AIC) will be applied for the purpose. With regression equations, p-values of regression coefficients tell you something about the goodness of fit of the independent variables. However, it says nothing about the complexity of the regression model applied, and it is easy to find a very complex model, that produces very good p-values. The trick is, thus, to find a regression model, that has the best trade-off between goodness and complexity. This is what the AIC will find out. The smaller the AIC of a model, the better the trade-off. In other words, each regression model suffers from information loss. The better the model, the better information loss is minimized. We will start by downloading and opening our data file, entitled "measured bodysurface", and stored in SpringerLink supplementary files, in our computer with SPSS statistical software installed. Then command.

Command:

click Analyze....click Generalized Linear Models....click Generalized Linear Models ....click Type of Model....click Scale Response....click Linear....click Response.... click Dependent Variable:...enter bodysurface measured....click Predictors....click Factors: ....enter gender, age, weight, height....click Model....click Main effects.... enter gender, age, weight, height....click Estimation....click Maximal likelihood estimate....click Statistics....click Log-Likelihood Function....click Full....click OK.

Goodness of Fit[a]

	Value	df	Value/df
Deviance	134551,397	5	26910,279
Scaled Deviance	90,000	5	
Pearson Chi-Square	134551,397	5	26910,279
Scaled Pearson Chi-Square	90,000	5	
Log Likelihood[b]	-456,650		
Akaike's Information Criterion (AIC)	1085,299		
Finite Sample Corrected AIC (AICC)	6073,299		
Bayesian Information Criterion (BIC)	1300,283		
Consistent AIC (CAIC)	1386,283		

Dependent Variable: bodysurface measured
Model: (Intercept), VAR00001, VAR00002, VAR00003, VAR00004

a. Information criteria are in small-is-better form.

b. The full log likelihood function is displayed and used in computing information criteria.

In the output sheets a goodness of fit is given. Particularly, the Akaike's Information Criterion is relevant.

Next, we will assess the goodness of fit of the kernel model. The same commands can be given, except for the three final commands, that should be....click Log-Likelihood Function....click Kernel....click OK.

The underneath goodness of fit table is given.

Goodness of Fit[a]

	Value	df	Value/df
Deviance	134551,397	5	26910,279
Scaled Deviance	90,000	5	
Pearson Chi-Square	134551,397	5	26910,279
Scaled Pearson Chi-Square	90,000	5	
Log Likelihood[b]	-373,945		
Akaike's Information Criterion (AIC)	919,890		
Finite Sample Corrected AIC (AICC)	5907,890		
Bayesian Information Criterion (BIC)	1134,874		
Consistent AIC (CAIC)	1220,874		

Dependent Variable: bodysurface measured
Model: (Intercept), VAR00001, VAR00002, VAR00003, VAR00004

a. Information criteria are in small-is-better form.

b. The kernel of the log likelihood function is displayed and used in computing information criteria.

The AICs of the two models are respectively 1085 and 920, difference 165. This means, that the probability of the kernel regression model to minimize information loss is e $^{(165)/2}$ times less worse than the full loglikelihood model is. Obviously, the kernel model performs endlessly better.

1.4   Conclusion

The Kernel regression method is more sensitive than traditional linear regression. However, kernel ridge regression is still parsimonious, because it has distinct advantages over linear and even polynomial regression. First, kernel regression is a discretization model. By the add-up sum of Gaussians continuous variables are converted into discrete ones, otherwise discretized ones. And like any discretization model it thus suffers from data *overfit*. This could be corrected for by regularization (correcting discretized variables for overfitting). With the recent kernel ridge regression, which is the main objective of this edition, ridge regularization is used for the purpose. Additional advantages of kernel ridge regression will be soon covered.

1.5 References

All of the chapters of the current edition start with a brief review of the traditional analytic method of the different regression methods prior to the review of the relevant kernel regression methods. For the purpose, generally, data examples are used from the recent edition "Regression Analyses in Clinical Research for Starters and 2nd Levelers 2nd Edition, Springer Heidelberg Germany 2021", by the same authors. For a better understanding of differences between traditional and kernel ridge regressions, readers may benefit from the study of this edition first.

To readers requesting still more background, theoretical and mathematical information of computations given, several textbooks complementary to the current production and written by the same authors are available: Statistics applied to clinical studies 5th edition, 2012, Machine learning in medicine a complete overview 2nd edition, 2020, SPSS for starters and 2nd levelers 2nd edition, 2015, Clinical data analysis on a pocket calculator 2nd edition, 2016, Understanding clinical data analysis from published research, 2016, all of them edited by Springer Heidelberg Germany.

Chapter 2
Kernel Ridge Regression (KRR)

Abstract Kernel regression is more sensitive than traditional ordinary least squares regression, but is a *discretization model*. By the add-up sum of Gaussians, continuous variables are converted into discrete ones, otherwise discretized ones.

Another problem is that of increasing mathematical complexity with multidimensional data. However, the *kernel trick* is an efficient and less computationally-intensive way to transform data into high dimensions. A third problem, is that of data overfitting. It can, however, be corrected by regularization, where regression coefficients (b-values) are penalized to a lower level according to $b_{ridge} = b/(1 + \lambda)$ where λ = shrinking factor.

Keywords Kernel ridge regression · Discretization model · Kernel trick · Overfitting · Ridge regularization

2.1 Summary

Three problems with kernel regression have to be accounted, and have, indeed, have been accounted for by the recent kernel ridge methodology.

1. Kernel regression is more sensitive than traditional ordinary least squares regression, but as shown in the previous chapter it is a *discretization model*. By the add-up sum of Gaussians, continuous variables are converted into discrete ones, otherwise discretized ones.
2. Another problem is that of increasing mathematical complexity with multidimensional data. However, the *kernel trick* is an efficient and reduced computationally-intensive way to transform data into high dimensions, and it offers a wonderful solution to the problem.
3. A third problem, is that of *overfitting*, i.e., data patterns wider than compatible with random sampling. It can, however, be corrected with some kind of data regularization, where regression coefficients (b-values) are penalized to a lower level. Ridge, lasso, elastic net regularization are accepted manners. And with

T. J. Cleophas, A. H. Zwinderman, *Kernel Ridge Regression in Clinical Research*,
https://doi.org/10.1007/978-3-031-10717-7_2

kernel regressions generally ridge regularization can be successfully used. It reduces b-values according to $b_{ridge} = b/(1 + \lambda)$ where λ = shrinking factor.

2.2   History of Kernel Ridge Regression

When was kernel ridge regression invented? The theory was first introduced by AE Hoerl and RW Kennard, computer scientists from the University of Delaware, the state at the mouth of the river Hudson first inhabited by former Dutchmen. The authors' 1970 seminal paper in Technometrics (1970; 8: 27–51) was entitled "Ridge regression biased estimation of nonorthogonal problems". The paper was the result of ten years of prior research into the field of ridge analysis. Six years earlier Nadaraya from Tsibilisi State University Georgia, and Watson from Victoria University Australia had independently published papers of a novel non-parametric regression method based on y-values expressed in the form of multiple Gaussian curves rather than histograms, and had called it kernel regression. This method was more powerful than ordinary least square regression, but results tended to be overfitted, because data spread was wider than compatible with random sampling.

2.3   Kernel Density Modeling

Traditional regression analysis assesses the effect of one or more predictor variables on an outcome variable. The outcome variable is assumed to be parametric, i.e., although the x-values are assumed to be without error, the y-values have uncertainties in the form of identical normal curves (Gaussian curves as demonstrated directly below).

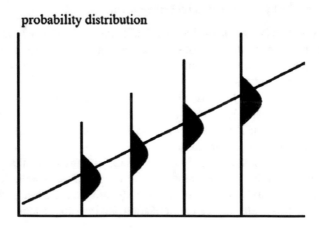

It means, that the uncertainty of the y-values is not expressed in the form of histograms fitted to Gaussian curves, like with traditional ordinary least square regressions, but rather, as the add-up sum of multiple identical Gaussian curves. The underneath graph gives an example.

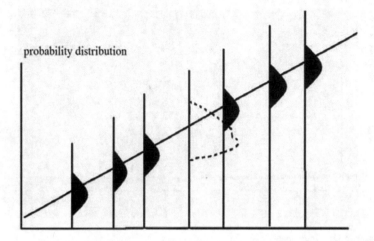

The dotted line is the add-up sum of the separate Gaussians with different intervals. This procedure will generally produce a better data fit, than traditional linear regression modeling. Instead of the areas under the curve of the add-up of identical Gaussian curves in the form of a large Gaussian-like curve, alternative and sometimes better fit mathematical function models can be used for fitting,

cosinal models,
additive-chi-square models,
chi-square models,
polynomial models,
laplacian models,
sigmoidal models,
radial basis function models.

They can for almost any data provide the very best fit data modeling. IBM's SPSS statistical software 2022 version 28.0.1.0 does, actually, offer eight of these different kernel density models, that should be helpful for finding the best fit kernel ridge model for your data. Underneath several graphs of a few of these mathematical models have been drawn.

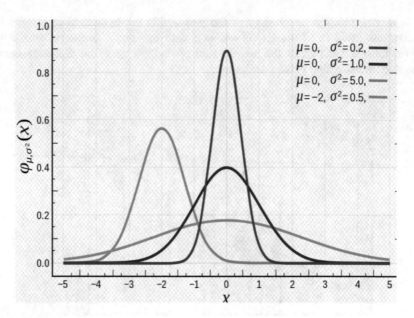

Four Gaussian Distributions

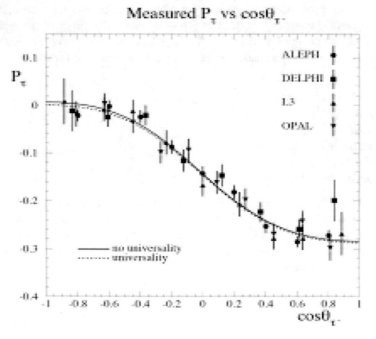

Cosinal Distribution

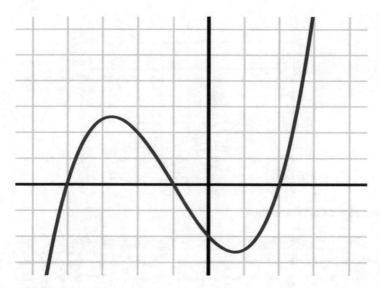

3rd Order Polynomial Distribution

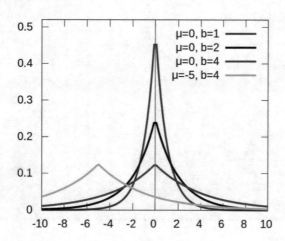

Multiple Laplace Distributions

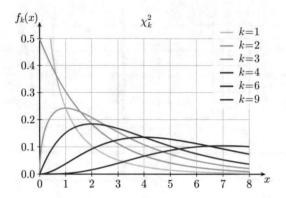

Additive-chi2 and Chi2 distributions

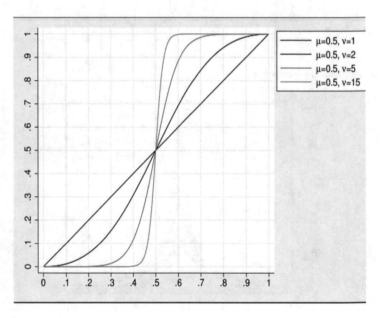

Sigmoid distributions

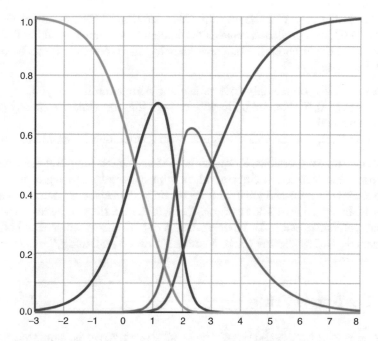

RBF (radial basis function) distributions

Much of published data on kernel ridge regressions is in users' guides, proceedings of symposia and conferences, and is, thus, of a rather theoretical and
preliminary nature.

Most kernel ridge regression (KRR) papers and communications are.

1. about basic sciences like nature, biology, chemistry, and physics,
2. about computer, data, and science,
3. about econo-/sociometrics.

Studies of clinical research are, however, virtually lacking so far. The current edition
assesses the clinical data analyses of kernel ridge regressions against traditional
ordinary least squares regressions as well as other para- and nonpara-metric regression models. Given the obvious advantages of reduced mathematical complexity by
the kernel trick, the easy applicability of multiple different kernel density models,
and the ridge protection from overfitting, the novel method should perform excellent,
particularly when alternative methods fail.

We should add, that the kernel ridge module in SPSS does not present computed
results tested with null hypothesis tests including p-values for assessing type I errors.
Instead, R-Square values of prediction are used as the main criterion for the power of
a kernel ridge regression. These R Square values are interpreted similarly as the R
Square values of traditional linear regressions. If R square = 0, then there will no
correlation between the dependent and independent variables. If R Square = 1000,
then there will be a 100% correlation. You will be sure about the outcome, if you

know the predictor(s). What about values between 0 and 1? For example, with an R Square of 0,63, the correlation between the predictors and outcome will be 0,63. You will be 63% certain about the outcome knowing the predictors. It can be demonstrated, that with

R Squares <0,25, you are only 25% certain, the correlation is very poor.
R Squares 0,25-0,50, you are up to 50% certain, this correlation is already considered reasonable.
R Squares >0,50, you are over 50% certain, this correlation is considered strong.

With kernel ridge regressions, R Squares have the same meaning. But there are some differences. Particularly, R Square values with a kernel ridge regression can be negative. How is it possible that with kernel ridge regression R square values can, obviously, be negative!!! The answer is, that with kernel ridge regression a negative R square value is possible for models where a data fit that is worse than horizontal (R Square = 0). See also the Chap. 5 entitled "Some Terminology".

2.4   The Kernel Trick

(Based on "What is the kernel trick, Zhang F, Nov 11 2018, Medium.Com").

Support vector machines can be used as a simplified cluster program for two dimensional data, that does not apply all of the observations in a dataset but rather the difficult ones lying close to the separation lines (see graph underneath). In the graph below, we notice that there are two classes of observations: the bullets and the crosses. There are several ways to separate these two classes as shown in the graph on the left. However, we want to find the "best" separation line or, with >2 dimensions (Ds). hyperplane that could maximize the margin between these two classes, which means that the distance between the hyperplane and the nearest data points on each side is the largest. Depending on which side of the hyperplane a new data point locates, we could assign a class to the new observation.

It sounds simple in the example given. However, not all of the data in a 2D plane are linearly separable. In fact, in the real world, almost all of the data are randomly distributed, and this makes it often hard to separate different classes linearly, at least in 2D.

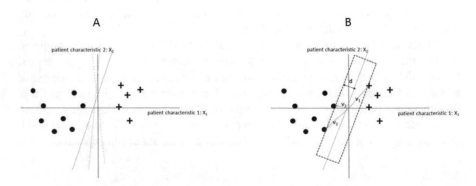

As you can see in the underneath picture, if we find a way to map the data from 2-dimensional space to 3-dimensional space, we will be able to find a decision plane that clearly divides the data into two different classes. My first thought of this data transformation process is to map all of the data points to a higher dimension (in this case, 3 dimensions), find the boundary, and make the classification.

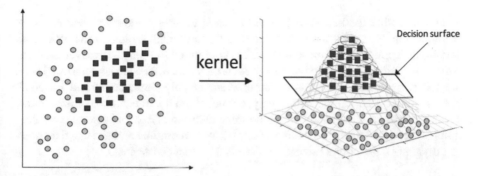

That sounds alright. However, when there are more and more dimensions, computations within that multidimensional space become more and more complex. This is, when the kernel trick comes in. It allows us to operate in the original basis space without computing the coordinates of the data in a higher dimensional space.

Let's look at an example to see how it works.

$$\mathbf{x} = (x_1, x_2, x_3)^T$$
$$\mathbf{y} = (y_1, y_2, y_3)^T$$

Here x and y are two data points in 3 dimensions. T means number of transfers, i.e., values of x in another dimension. Let's assume that we need to map x and y not in a 3 but in a 9-dimensional space. We need to do the following calculations to get the final results, which are just dimensionless quantities. The computational complexity, in this case, is $O(n^2)$.

$$\phi(\mathbf{x}) = \left(x_1^2, x_1 x_2, x_1 x_3, x_2 x_1, x_2^2, x_2 x_3, x_3 x_1, x_3 x_2, x_3^2\right)^T$$
$$\phi(\mathbf{y}) = \left(y_1^2, y_1 y_2, y_1 y_3, y_2 y_1, y_2^2, y_2 y_3, y_3 y_1, y_3 y_2, y_3^2\right)^T$$

$$\phi(\mathbf{x})^T \phi(\mathbf{y}) = \sum_{i,j=1}^{3} x_i x_j y_i y_j$$

However, if we use the kernel function, which is denoted as k (x, y), instead of doing the complicated computations in the 9-dimensional space, we will reach the same result within the 3-dimensional space by calculating the dot product of x -transpose and y. The computational complexity, in this case, is only O(n).

$$k(\mathbf{x}, \mathbf{y}) = \left(\mathbf{x}^T \mathbf{y}\right)^2$$
$$= (x_1 y_1 + x_2 y_2 + x_3 y_3)^2$$
$$= \sum_{i,j=1}^{3} x_i x_j y_i y_j$$

In essence, what the kernel trick does for us is to offer a more efficient and less computationally intensive way to transform data into higher dimensions. With that saying, the application of the kernel trick is not limited to a single dimensional step. Any computations involving the dot products (x, y) can utilize the kernel trick. There are different kernels. The most popular ones are the polynomial kernel and the radial basis function (RBF) kernel. Intuitively, the polynomial kernel looks not only at the given features of input samples to determine their similarity, but also at combinations of these (just like in the above example). With n original features and d degrees of polynomial, the polynomial kernel yields n ^{d} expanded features.

$$k(\mathbf{x}, \mathbf{y}) = \left(\mathbf{x}^T \mathbf{y} + 1\right)^d$$

This is the format of the polynomial kernel. The RBF (radial basis function) kernel is also called the Gaussian kernel. There is an infinite number of dimensions in the feature space, because it can be expanded by a Taylor Series (Cleophas, Zwinderman, Regression Analysis in Medical Research, edited by Springer Heidelberg Germany: 2021, pp. 267, 268). In the format below, The γ parameter defines how much influence a single training example has. The larger it is, the closer other examples must be affected.

$$k(\mathbf{x}, \mathbf{y}) = e^{-\gamma \|\mathbf{x} - \mathbf{y}\|^2}, \gamma > 0$$

This is the format of the RBF kernel. There are different options for the kernel functions in the SPSS version kernel ridge menu. You can even build a custom kernel if needed.

Finally, the kernel trick sounds like a "perfect" plan. However, one critical thing to keep in mind is that, when we map data to a higher dimension, there are chances that we may further overfit the model. Thus choosing the right kernel function (including the right parameters) and the right regularization are of great importance.

This may sound pretty much Greek to non-mathematicians, but for those with little affection to calculus: you may skip these pages and yet start the self-assessment analyses abundantly given in all of the following chapters without loss of understanding how kernel ridge regression does work in practice. In contrast, for those with mathematical talent, a more in depth description is in the Chap. 27 entitled "The Kernel Trick".

2.5 Ridge Regularization

A third problem, is that of *overfitting*, i.e., data patterns wider than compatible with random sampling. It means the production of an analysis that corresponds too closely or exactly to a particular dataset, and, therefore, fails to fit to additional data or predict future observations reliably. Overfitting can nonetheless be easily be corrected with some kind of data regularization, where regression coefficients (b-values) are penalized to a lower level. Ridge, lasso, elastic net regularization are accepted manners. And with kernel regressions generally ridge regularization can produce very well fitting data patterns. It reduces b-values according to $b_{ridge} = b/(1 + \lambda)$ where λ = shrinking factor.

2.6 Conclusion

Kernel regression is more sensitive than traditional ordinary least squares regression, but as shown in the previous chapter it is a discretization model. By the add-up sum of Gaussians continuous variables are converted into discrete ones, otherwise discretized ones. Another problem is that of increasing mathematical complexity with multidimensional data. However, the *kernel trick* offers a wonderful solution to the problem. A third problem, is that of overfitting, i.e., data patterns wider than compatible with random sampling. It can, however, be corrected with some kind of data regularization, where regression coefficients (b-values) are penalized to a lower level. Ridge, lasso, elastic net regularization are accepted manners. And with kernel regressions generally ridge regularization can be successfully used. It reduces b-values according to $b_{ridge} = b/(1 + \lambda)$ where λ = shrinking factor.

2.7 References

All of the chapters of the current edition start with a brief review of the traditional analytic method of the different regression methods prior to the review of the relevant kernel ridge regression method. For the purpose, generally, data examples are used from the recent edition "Regression Analyses in Clinical Research for Starters and 2nd Levelers 2nd Edition, Springer Heidelberg Germany 2021", by the same authors. For a better understanding of differences between traditional and kernel ridge regressions, readers may benefit from the study of this edition first.

To readers requesting still more background, theoretical and mathematical information of computations given, several textbooks complementary to the current production and written by the same authors are available: Statistics applied to

clinical studies 5th edition, 2012, Machine learning in medicine a complete overview 2nd edition, 2020, SPSS for starters and 2nd levelers 2nd edition, 2015, Clinical data analysis on a pocket calculator 2nd edition, 2016, Understanding clinical data analysis from published research, 2016, all of them edited by Springer Heidelberg Germany.

Chapter 3
Optimal Scaling vs Kernel Ridge Regression

Abstract Optimal scaling is a possibility to improve the correlation between predictors and outcomes. Kernel ridge regression provides optimally fit correlations, and performs even better than optimal scaling for the purpose of optimized predictive modeling.

The traditional R Square values of the scale 1–3 models were respectively

scale 1 R Square = 0,277 = 27,7% certainty about the outcome
scale 2 R Square = 0,281 = 28,1% certainty about the outcome
scale 3 R Square = 0,380 = 38,0% certainty about the outcome.

The kernel ridge R Square values of the scale 1–3 models were respectively

scale 1 R Square = 0,455 = 45,5% certainty about the outcome
scale 2 R Square = 0,808 = 80,8% certainty about the outcome
scale 3 R Square = 0,962 = 96,2% certainty about the outcome.

Keywords Optimal scaling · Kernel ridge regression · R Square values

3.1 Summary

The effect of linear predictor on continuous outcome was assessed using different scales of the predictor values.

3.1.1 Summaries of the Traditional Linear Regressions

In summary the traditional R Square values of the scale 1–3 models were respectively

Supplementary Information The online version contains supplementary material available at [https://doi.org/10.1007/978-3-031-10717-7_3].

scale 1 R Square = 0,277 = 27,7% certainty about the outcome
scale 2 R Square = 0,281 = 28,1% certainty about the outcome
scale 3 R Square = 0,380 = 38,0% certainty about the outcome.

The p-values of the traditional linear regressions of the three scale models were respectively

scale 1 p-value = 0,079
scale 2 p-value = 0,076
scale 3 p-value = 0,033.

3.1.2   Summaries of the Kernel Ridge Regressions

The **scale 1** kernel ridge regressions produced four kernel density models with R Squares over 0,250. The best fit R Square was provided by the polynomial kernel density model, 0,455.

The **scale 2** kernel ridge regressions produced five kernel density models with R Squares over 0,250. The best fit R Square was provided by the polynomial kernel density model, 0,808.

The **scale 3** kernel ridge regressions produced six kernel density models with R Squares over 0,250. The best fit R Square was provided by the polynomial kernel density model, 0,962.

3.1.3   In Conclusion

Optimal scaling is a possibility to improve the correlation between predictors and outcomes. KRR provides optimally fit correlations, and performs even better than optimal scaling for the purpose of optimized predictive modeling.

3.2   Introduction

Linear regression analysis is often used for analyzing the effect of a predictor on an outcome variable. For example a continuous predictor variable can be scored on different outcome scales between 0 and 10. Often the relationship between outcome and predictor can be optimized in order for the predictor to better fit the outcome. In the data example of this chapter the predictor models produced different p-values with different scales. The scales 1, 2, 3 produced p-values of respectively 0,079, 0,076, and 0,033. Optimal scaling can be replaced or supplemented with other predictive statistical models. And in this edition kernel ridge regression (KRR) will be used for the purpose. KRR does not produce p-values but, instead, R square values where an R square of 1000 indicates, that the percentage of certainty given by the predictor of the outcome is 100%.

The underneath section gives tables and a graph of the above data example. For convenience the data file is in SpringerLink supplementary files, and is entitled "optimalscaling".

3.3 Optimal Scaling

A 12 patient data file with a single predictor and a single outcome was performed while using three different scales for data analysis.

Patients with the predictor values 0, 1, 5, 9 and 10 are missing. Instead of a scale of integers between 0 and 10, other scales are possible, e.g. a scale of two or four scores. Any scale used is, of course, arbitrary and can be replaced with another one.

Scale 1: 0, 1, 2, 3, 4, 5, 6, 7, 8, 9, 10.
Scale 2: 1, 2 (1 = (0 to 5); 2 = (5 to 10).
Scale 3: 1, 2, 3, 4 (1 = (0 to 2.5); 2 = (2.5 to 5); 3 = (5 to 7.5); 4 = (7.5 to 10)

The underneath table and graph show, that linear regressions of each scale produced different regression coefficients, t-values, and p-values, one result better than the other. With the scales 2 and 3 a gradual improvement of the R Square values, t-values, and p-values were observed. Optimal scaling is a method designed to optimize the statistical power of the relationship between the predictor and outcome variables. With optimal scaling the best fit predictor model produced a p-value of 0,033 at best. In contrast, with kernel ridge regressions different results were obtained.

outcome	scale1	scale2	scale3
2,00	3,00	1,00	1,00
2,00	5,00	1,00	1,00
3,00	7,00	1,00	2,00
3,00	8,00	1,00	2,00
4,00	1,00	1,00	2,00
4,00	4,00	1,00	2,00
6,00	6,00	2,00	3,00
6,00	8,00	2,00	3,00
7,00	4,00	2,00	3,00
7,00	8,00	2,00	3,00
9,00	8,50	2,00	4,00
9,00	9,00	2,00	4,00

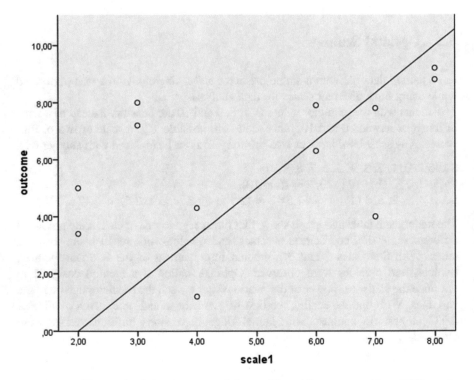

Above a 2D graph of data is assessed for traditional linear regression. Differently scaled x-variables can be used for the purpose of finding the optimal scale.

3.4 Traditional Regressions

Scale 1
Command:

Analyze....Menu....Regression....Linear....Dependent: enter outcome....Independent (s): enter scale 1....click OK.

In the output sheets are the underneath two tables.

Model Summary

Model	R	R Square	Adjusted R Square	Std. Error of the Estimate
1	,526[a]	,277	,205	2,26914

a. Predictors: (Constant), predictor

ANOVA[a]

Model		Sum of Squares	df	Mean Square	F	Sig.
1	Regression	19,739	1	19,739	3,834	,079[b]
	Residual	51,490	10	5,149		
	Total	71,229	11			

a. Dependent Variable: outcome

b. Predictors: (Constant), predictor

The above scale 1 produced an insignificant p-value of 0,079.
The underneath scale 2 analysis produced an insignificant p-value of 0,076.

Scale 2
Similar commands are given. The output is underneath.

Model Summary

Model	R	R Square	Adjusted R Square	Std. Error of the Estimate
1	,530[a]	,281	,209	2,26293

a. Predictors: (Constant), scale2

ANOVA[a]

Model		Sum of Squares	df	Mean Square	F	Sig.
1	Regression	20,021	1	20,021	3,910	,076[b]
	Residual	51,208	10	5,121		
	Total	71,229	11			

a. Dependent Variable: outcome

b. Predictors: (Constant), scale2

Scale 3

Similar commands are again given. The output is below.

Model Summary

Model	R	R Square	Adjusted R Square	Std. Error of the Estimate
1	,616[a]	,380	,318	2,10186

a. Predictors: (Constant), scale3

ANOVA[a]

Model		Sum of Squares	df	Mean Square	F	Sig.
1	Regression	27,051	1	27,051	6,123	,033[b]
	Residual	44,178	10	4,418		
	Total	71,229	11			

a. Dependent Variable: outcome

b. Predictors: (Constant), scale3

Finally the scale 3 analysis produced a significant p-value of 0,033.

In summary the traditional R Square values of the scale 1–3 models were respectively.

scale 1 R Square = 0,277
scale 2 R Square = 0,281
scale 3 R Square = 0,380

The p-values of the traditional linear regressions of the three scale models were respectively.

scale 1 p-value = 0,079
scale 2 p-value = 0,076
scale 3 p-value = 0,033.

3.5 Kernel Ridge Regressions Scale 1

We will perform kernel ridge regression on the same data, and will see, if a better fit result can be obtained by any of the scales. Remember that the novel revised version of SPSS 2022, 18.0.1.0 is required in your computer.

Command:

Menu....Analyze....Regression....Kernel Ridge Regression....Dependent: Outcome....Independent(s): Scale1....click Options: mark Observed vs Predicted.... mark Predicted values....click Continue....click Linear....click OK.

Model Summary[a,b]

Kernel	Alpha	R Square
Linear	1,000	,172

a. Dependent Variable:
   outcome

b. Model: scale1

In order to find better fit kernel ridge regressions seven more kernel density models were assessed using similar commands.

Note, that negative kernel ridge R Square values are frequently observed. The phenomenon is explained in the Chap. 5 entitled "Some Terminology".

Model Summary[a,b]

Kernel	Alpha	R Square
Additive_chi2	1,000	,336

a. Dependent Variable: outcome

b. Model: scale1

Model Summary[a,b]

Kernel	Alpha	Gamma	R Square
Chi2	1,000	1,000	,144

a. Dependent Variable: outcome

b. Model: scale1

Model Summary[a,b]

Kernel	Alpha	R Square
Cosine	1,000	-,027

a. Dependent Variable:
   outcome

b. Model: scale1

Model Summary[a,b]

Kernel	Alpha	Gamma	R Square
Laplacian	1,000	1,000	,304

a. Dependent Variable: outcome

b. Model: scale1

Model Summary[a,b]

Kernel	Alpha	Gamma	Coef0	Degree	R Square
Polynomial	1,000	1,000	1,000	3,000	,455

a. Dependent Variable: outcome

b. Model: scale1

Model Summary[a,b]

Kernel	Alpha	Gamma	R Square
RBF	1,000	1,000	,275

a. Dependent Variable: outcome

b. Model: scale1

Model Summary[a,b]

Kernel	Alpha	Gamma	Coef0	R Square
Sigmoid	1,000	1,000	1,000	-,028

a. Dependent Variable: outcome

b. Model: scale1

The above **scale 1** kernel ridge regressions produced four kernel density models with R Squares over 0,250. The best fit R Square was provided by the polynomial kernel density model, 0,455.

3.6 Kernel Ridge Regressions Scale 2

The underneath **scale 2** kernel ridge regression were subsequently assessed using similar commands.

Model Summary[a,b]

Kernel	Alpha	R Square
Linear	1,000	,772

a. Dependent Variable: outcome

b. Model: scale2

Model Summary[a,b]

Kernel	Alpha	R Square
Additive_chi2	1,000	-3,879

a. Dependent Variable: outcome

b. Model: scale2

Model Summary[a,b]

Kernel	Alpha	Gamma	R Square
Chi2	1,000	1,000	,662

a. Dependent Variable: outcome

b. Model: scale2

Model Summary[a,b]

Kernel	Alpha	R Square
Cosine	1,000	-,027

a. Dependent Variable: outcome

b. Model: scale2

Model Summary[a,b]

Kernel	Alpha	Gamma	R Square
Laplacian	1,000	1,000	,719

a. Dependent Variable: outcome

b. Model: scale2

Model Summary[a,b]

Kernel	Alpha	Gamma	Coef0	Degree	R Square
Polynomial	1,000	1,000	1,000	3,000	,808

a. Dependent Variable: outcome

b. Model: scale2

Model Summary[a,b]

Kernel	Alpha	Gamma	R Square
RBF	1,000	1,000	,719

a. Dependent Variable: outcome

b. Model: scale2

Model Summary[a,b]

Kernel	Alpha	Gamma	Coef0	R Square
Sigmoid	1,000	1,000	1,000	-,133

a. Dependent Variable: outcome

b. Model: scale2

The **scale 2** kernel ridge regressions produced five kernel density models with R Squares over 0,250. The best fit R Square was provided by the polynomial kernel density model, 0,808.

3.7   Kernel Ridge Regressions Scale 3

The underneath **scale 3** kernel ridge regressions were subsequently assessed using similar commands.

Model Summary[a,b]

Kernel	Alpha	R Square
Linear	1,000	,930

a. Dependent Variable: outcome

b. Model: scale3

Model Summary[a,b]

Kernel	Alpha	R Square
Additive_chi2	1,000	,669

a. Dependent Variable: outcome

b. Model: scale3

Model Summary[a,b]

Kernel	Alpha	Gamma	R Square
Chi2	1,000	1,000	,763

a. Dependent Variable: outcome

b. Model: scale3

Model Summary[a,b]

Kernel	Alpha	R Square
Cosine	1,000	-,027

a. Dependent Variable: outcome

b. Model: scale3

Model Summary[a,b]

Kernel	Alpha	Gamma	R Square
Laplacian	1,000	1,000	,741

a. Dependent Variable: outcome

b. Model: scale3

Model Summary[a,b]

Kernel	Alpha	Gamma	Coef0	Degree	R Square
Polynomial	1,000	1,000	1,000	3,000	,962

a. Dependent Variable: outcome

b. Model: scale3

Model Summary[a,b]

Kernel	Alpha	Gamma	R Square
RBF	1,000	1,000	,726

a. Dependent Variable: outcome

b. Model: scale3

Model Summary[a,b]

Kernel	Alpha	Gamma	Coef0	R Square
Sigmoid	1,000	1,000	1,000	-,075

a. Dependent Variable: outcome

b. Model: scale3

The above **scale 3** kernel ridge regressions produced six kernel density models with R Squares over 0,250. The best fit R Square was provided by the polynomial kernel density model, 0,962.

3.8   Conclusion

Optimal scaling is a traditional method for optimizing the data fit of regression models. In the example the p-values of testing improved from 0.,079 to 0,033. The traditional R-square values improved from 0,277 to 0,380. With kernel ridge regression much better fit results were obtained.

3.8.1   Summary of the Traditional R Square Values of the Scales 1–3 Models

scale 1 R Square = 0,277
scale 2 R Square = 0,281
scale 3 R Square = 0,380

The p-values of the traditional linear regressions of the three scale models were respectively

scale 1 p-value = 0,079
scale 2 p-value = 0,076
scale 3 p-value = 0,033.

3.8.2   Summary of Kernel Ridge Regressions (KRR)

R square polynomial KRR.

The **scale 1** kernel ridge regressions produced four kernel density models with R Squares over 0,250. The best fit R Square was provided by the polynomial kernel density model, 0,455.

The **scale 2** kernel ridge regressions produced five kernel density models with R Squares over 0,250. The best fit R Square was provided by the polynomial kernel density model, 0,808.

The **scale 3** kernel ridge regressions produced six kernel density models with R Squares over 0,250. The best fit R Square was provided by the polynomial kernel density model, 0,962.

3.9   References

All of the chapters of the current edition start with a brief review of the traditional analytic method of the different regression methods prior to the review of the relevant kernel ridge regression method. For the purpose, generally, data examples are used from the recent edition "Regression Analyses in Clinical Research for Starters and 2nd Levelers 2nd Edition, Springer Heidelberg Germany 2021", by the same authors. For a better understanding of differences between traditional and kernel (ridge) regressions, readers may benefit from the study of this edition first.

To readers requesting still more background, theoretical and mathematical information of computations given, several textbooks complementary to the current production and written by the same authors are available: Statistics applied to clinical studies 5th edition, 2012, Machine learning in medicine a complete overview 2nd edition, 2020, SPSS for starters and 2nd levelers 2nd edition, 2015, Clinical data analysis on a pocket calculator 2nd edition, 2016, Understanding clinical data analysis from published research, 2016, all of them edited by Springer Heidelberg Germany.

Chapter 4
Examples of Published Kernel Ridge Regressions So Far

Abstract The theory of kernel ridge regressions was first introduced by AE Hoerl and RW Kennard from Delaware. The authors seminal paper in Technometrics (1970; 8: 27–51) summarized several advantages: the kernel trick for reduced arithmetic complexity, the estimation of uncertainty through Gaussians instead of histograms, the corrected data-overfit through ridge regularization, the availability of multiple alternative kernel density models for improved data fit. A brief search of competitive kernel ridge regression publications revealed one genomic study, one facial surgery study, one genetic study. The remainder of publications involved chemistry studies, econometry studies, and studies from basic sciences like nature, biology and physics. Also climate studies were observed.

Keywords Kernel ridge regression · Ridge regularization · Kernel density modeling

4.1 Summary

The theory of kernel ridge regressions was first introduced by AE Hoerl and RW Kennard, computer scientists from the University of Delaware. The authors' 1970 seminal paper in Technometrics (1970; 8: 27–51) was entitled "Ridge regression based estimation of nonorthogonal problems".

Theoretical advantages include (1) kernel trick for reduced arithmetic complexity, (2) estimation of uncertainty through Gaussians instead of histograms, (3) corrected data-overfit through ridge regularization, (4) availability of multiple alternative kernel density models for improved data fit.

A brief search of competitive kernel ridge regression publications revealed

one genomic study,
one facial surgery study,
one genetic study.

The remainder of publications involved chemistry studies, econometry studies, and studies from basic sciences like nature, biology and physics. Also climate studies were observed.

4.2   Introduction

IBM (International Business Machines) company has published in its SPSS statistical software 2022 version 18.01.0 update a very important novel regression method entitled Kernel Ridge Regression (KRR). It is an extension of the currently available regression methods, and is suitable for pattern recognition in high dimensional data particularly when alternative methods fail. Its theoretical advantages are plenty and include the

1. kernel trick for reduced arithmetic complexity,
2. estimation of uncertainty through Gaussians instead of histograms
3. corrected data-overfit through ridge regularization
4. availability of eight alternative kernel density models for improved data fit

A very exciting and wide array of research has already been published by major disciplines, like studies in quantum mechanics and nuclear physics, studies of molecular affinity/dynamics, atomisation energy studies, but also forecasting economics studies, IoT (internet of things) studies for e-networks, plant stress response studies, big data streaming studies etc. In contrast, it is virtually unused in clinical research.

4.3   History of Kernel Ridge Regression

When was kernel ridge regression invented? The theory was first introduced by AE Hoerl and RW Kennard, computer scientists from the University of Delaware, the state at the mouth of the river Hudson first inhabited by former Dutchmen. The authors' 1970 seminal paper in Technometrics (1970; 8: 27–51) was entitled "Ridge regression based estimation of nonorthogonal problems". The paper was the result of ten years of prior research into the field of ridge analysis. Six years earlier Nadaraya from Tsibilisi State University Georgia, and Watson from Victoria University Australia had independently published papers of a novel non-parametric regression method based on y-values expressed in the form of multiple Gaussian curves rather than histograms, and had called it kernel regression. This method was more powerful than ordinary least square regression, but results tended to be overfitted, because data spread was wider than compatible with random sampling.

Support vector machines is a simplified cluster program that does not apply all of the observations in a dataset but rather the difficult ones lying close to the separation lines. In the graph below, we notice that, in a 2D plane, there are two classes of observations: the dots and the crosses.

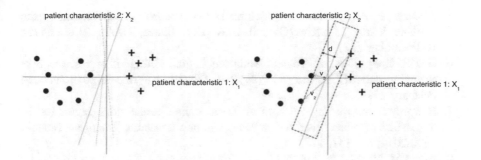

There are numerous ways to separate these two classes as shown in the graph on the left. However, we want to find exactly the "very best" separation line (or with >2 dimensions) hyperplane that could maximize the margin between these two classes, which means that the distance between the hyperplane and the nearest data points on each side is the largest. Depending on which side of the hyperplane a new data point is located, we could assign a class to the new observation.

It sounds and looks simple in the above example. However, not all of the data are linearly separable. In fact, in the real world, almost all of the data are randomly distributed, which makes it hard to separate different classes linearly. The kernel trick is helpful for the purpose (Chap. 2), but kernel assessment applies the means of add-up scores, and is thus easily overfitted, i.e., data are too close to functions and functions can, therefore, not be used for making predictions. Overfitting can nonetheless be easily corrected for with some kind of data regularization, where regression coefficients (b-values) are penalized to a lower level. Ridge, lasso, elastic net regularization are accepted manners. And with kernel regressions generally ridge regularization can be successfully used. It reduces b-values according to $b_{ridge} = b/(1 + \lambda)$ where $\lambda =$ shrinking factor.

4.4   A Brief Search of Kernel Ridge Regression Publications So Far

1. Safari M, Arashloo S. Kernel ridge regression model for sediment transport in open channels. Neural Computing & Applications 2021; 33: 11255–11271.
2. Alalami M, Maalouf M, El-Fouly T. Wind speed forecasting using kernel ridge regression with different time horizons. In book: Theory and applications of time series analysis selected from ITISE (International conference on time series and forecasting) 2–10; pp. 191–203. Doi: 10.1007/978-3-030-56219-9-13.
3. Exterkate P, Groenen P, et al. Nonlinear forecasting with many predictors using kernel ridge regression. in: Tinbergen Institution Discussion Paper Rotterdam, 2011: paper 007/4.
4. Ferre G, Haut T, Barros K. Learning energies using localized graph kernels. arXiv (Cornell University, Ithaca, USA): 1612. 00193v2, 25 April 2017.

5. Machaka P, Aiayi O et al. Modelling DDoS Attacks in IoT Networks using machine learning. arXiv (Cornell University, Ithaca, USA): 2112.0547v1, 10 December 2021.
6. Gao F, Song X et al. Toward budgeted Online kernel ridge regression on streaming data. IEEE Access PP(99) 1–1. Doi:10.1109/Access.2019.2900014, February 2019.
7. Haworth J, Shawe-Taylore J, et al. Local online kernel ridge regression for forecasting of urban travel times. Transportation Research: Emerging Technologies 2014; 46: 151–178.
8. Asaari M, Mertens S, et al. Analysis of plant stress response using hyperspectral imaging and kernel ridge regression. Proceedings of 11th International Conference on Robotics, Vision, Signal processing and Power Applications Conference Paper, 1 January 2022.
9. Montesinos A, Montesinos O. A guide for generalized regression methods for genomic-enabled prediction. Heredity 2021; 126: 577–596.
10. Zheng A and Sun L. Regularized least square kernel regression for streaming data. Communications in Mathematical Sciences. 2021; 19: 1533–1548.
11. Pan B, Xia J, et al. Incremental kernel ridge regression for prediction of soft tissue deformations after maxillofacial surgery. Med Image Comput Assist Interv 2012; 15: 99–106.
12. Shen C, Ding y, et al. Multivariate information fusion with fast kernel learning to kernel ridge regression in predicting lncRNA-protein interactions. Front Genet 2019; 15 January, Doi.org/10.3389/fgene.2018.00716.

The brief search of competitive kernel ridge regression publications revealed.

one genomic study (study 9),
one facial surgery study (study 11),
one genetic study (study 12).

The remainder of the publications involved chemistry studies, econometry studies, and studies from basic sciences like nature, biology and physics. Also climate studies were involved.

4.5   Courses Where the Upcoming Edition "Kernel Ridge Regression in Clinical Research" Will Be Used

1. Pharmacovigilance Semesters 2022, University Lyon 1 Faculty Laennec, (Professor Cleophas Member Scientific Committee).
2. Ongoing Big Statistics Courses 2022 for Omics Research (Genome, Proteonome, Metabolome) at the Amsterdam Universities, (Professor Dr. A.H.Zwinderman Member Scientific Committee and Lecturer).

3. Continued Medical Education Courses and Practical
   Biostatistics 2022 at the University of Amsterdam AMC and the Free University
   of Amsterdam (Professor Dr. A.H. Zwinderman session chair).
4. Global Annual Meetings of the Drug Information Association (DIA) March
   29–31 Brussels (Professor Dr.T.J.Cleophas session chair).
5. Yearly Statistics Modules and Exams of the Masters Diploma in Pharmaceutical
   Medicine (December 12–14, 2022 Lyon Claude Bernard University (Member
   Scientific Committee Professor Dr. T.J. Cleophas and Lecturer Professor Dr. A.
   H.Zwinderman)).
6. KNIME (Konstanz Information Miner) Yearly Summit, February 28–March
   4, 2022, online.

4.6   Conclusion

The theory of kernel ridge regressions was first introduced by AE Hoerl and RW
Kennard, computer scientists from the University of Delaware. The authors' 1970
seminal paper in Technometrics (1970; 8: 27–51) was entitled "Ridge regression
based estimation of nonorthogonal problems".

Theoretical advantages include (1) kernel trick for reduced arithmetic complexity,
(2) estimation of uncertainty through Gaussians instead of histograms, (3) corrected
data-overfit through ridge regularization, (4) availability of multiple alternative
kernel density models for improved data fit.

A brief search of competitive kernel ridge regression publications revealed one
genomic study, one facial surgery study, one genetic study. The remainder of
publications involved chemistry studies, econometry studies, and studies from
basic sciences like nature, biology and physics. Also climate studies were observed.

4.7   References

All of the chapters of the current edition start with a brief review of the traditional
analytic method of the different regression methods prior to the review of the
relevant kernel ridge regression method. For the purpose, generally, data examples
are used from the recent edition "Regression Analyses in Clinical Research for
Starters and 2nd Levelers 2nd Edition, Springer Heidelberg Germany 2021", by
the same authors. For a better understanding of differences between traditional and
kernel regressions, readers may benefit from the study of this edition first.

To readers requesting still more background, theoretical and mathematical information of computations given, several textbooks complementary to the current
production and written by the same authors are available: Statistics applied to

clinical studies 5th edition, 2012, Machine learning in medicine a complete overview 2nd edition, 2020, SPSS for starters and 2nd levelers 2nd edition, 2015, Clinical data analysis on a pocket calculator 2nd edition, 2016, Understanding clinical data analysis from published research, 2016, all of them edited by Springer Heidelberg Germany.

Chapter 5
Some Terminology

Abstract Twenty-four terms particularly often used in connection with kernel ridge regressions are given. Many more new terminologies are applied in the current edition, but most of them have been explained directly in the text.

Keywords Kernel ridge regression · Novel terminologies · Kernel trick · Ridge regularization · Kernel density modeling

5.1 Summary

Summary of Terms:

Exhausive searching
Graphical frequency distributions
Histograms and graphical frequency distributions vs mathematical frequency distributions
Interpretation of R Square values
IoT (internet of things)
Kernel density model
Kernel trick
Lacking p-values
Linear kernel model
Mathematical frequency distributions
Misnomer
Multicollinearity
Multivariate kernel ridge regression
Overfitting
Problems with R Square values
Pseudo Re Square values
R Square values that, with a kernel ridge regression, can be negative
Ridge Regularization
Scikit-learn or Sklearn

Study Stream
Stream Study
Test-Retest Reliability
Final note

5.2   Alphabetical Enumeration

E
Exhaustive Searching

A brute-force search method also known as "generate and test". It is a very general problem solving technique and algorithm, and it is typically applied with decision trees.

G
Graphical Frequency Distributions

Histograms like the one underneath are examples of graphical frequency distributions. They have individual data on the x-axis and "how often on the y-axis.

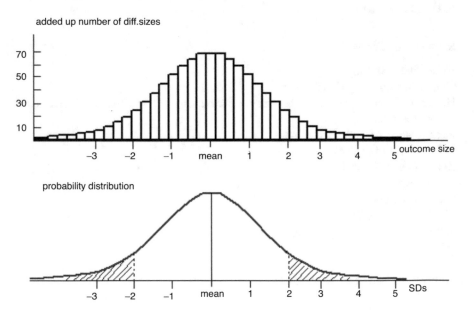

The upper graph is an example of a histogram of "normal data", otherwise called Gaussian frequency distribution. It has individual data on the x-axis and "how often on the y-axis. It is not adequate for testing statistical hypotheses. The lower graph has again individual data on the x-axis but on the y-axis bars have been replaced with a continuous line. Now it is impossible to determine from the graph the individual results, but, instead, important inferences can be made for example:

total auc (area under the curve) = 100% of the data of a trial
the auc left from the mean = 50% of data
the auc left from - 1 standard deviations = 15% of the data
the auc left from - 2 standard deviations = 2,5% of the data
the auc between −2 and + 2 standard deviations = 95% of the data.

How come that the above clock-like pattern has the above characteristics. It is simple a present of nature, and is so reproducible that it even can be used for making predictions reliable about future data.

H
Histograms and Graphical Frequency Distributions vs Mathematical Frequency distributions

Histograms usually have individual data on the x-axis and bars with "how often" on the y-axis. If the bars are replaced with a continuous line, then we will call the graph a frequency distribution. Frequency distributions often are close to having a mathematical function, and if so, they can, actually, be used as the core principle of regression analysis. Regression-analysis calculates best fit "line /exponential-curve / curvilinear-curve etc." (i.e. the curve with the shortest distance from the data) and it, then, tests, how far distant from the curve the data are. A significant correlation between y- and x-data means that the y data are closer to model than will happen with random sampling (i.e., by chance). The distance is usually statistically tested with simple null hypothesis tests like t-tests or analyses of variance. We should add and emphasize here that the model-principle is, generally, at the same time the largest limitation of statistical modeling, because often it is no use forcing nature into a statistical model.

I
Interpretation of R Square Values

R Square values 0 = no correlation between predictors and outcome
R Square 0–25% = a very poor correlation
R Square 25–50% = a reasonable correlation
R Square > 50% = a strong correlation
R Square 100% = a 100% correlation (certain about y, if knowing the x-values).

IoT (internet of things)

IoT (internet of things) describes and connects physical objects with sensors. It has processing ability, possesses software, and more technologies to connect and exchange data. Different platforms for IoT do exist, and they consist of individuals, their devices. They collect and analyze data of participating assets, locations, people. A great danger is fraud. And cyber attacks may very well originate from channels connecting IoT devices.

K
Kernel Density Model

In statistics, kernel density Modeling (kernel density estimation, KDE) is **a non-parametric way to estimate the mathematical probability density function of a random variable**. Kernel density estimation is a fundamental data smoothing problem where inferences about the population are made, based on a finite data sample.

Kernel Trick

A problem with kernel regression is of course the increasing mathematical complexity with high dimensional data. However, the kernel trick, explained in the Chap. 2, offers a wonderful solution.

L
Lacking P-Values

The kernel ridge module in SPSS statistical software does not present p-values for assessing type errors of finding effects where there are none. Instead, R Square values of prediction are used as the main outcome criterion for kernel ridge regressions.

Linear Kernel Model

In the recent past only the linear kernel model. It means that uncertainty of the y-values is not expressed in the form of histograms fitted to Gaussian curves like with traditional ordinary least square regressions, but rather as the add-up of sums of multiple Gaussian curves. This will generally produce a better data fit. Instead of the areas under the curve (AUCs) of Gaussian curves, other mathematical models can be applied, for example, cosinal, chi-square, polynomial, laplacian, sigmoidal, radial basis function curves. They can, for some data, provide even better fit data models. IBM's SPSS statistical software 2022 version 18 0.1.0 does actually offer 8 different kernel density models for the purpose.

M
Mathematical Frequency Distributions

Underneath various examples are given of mathematical functions that have been used for making predictions about observed data distribution that are slightly different but still so close to the mathematical models that they can be used for predictive purposes. Examples applied with kernel ridge regressions are in the next four pages.

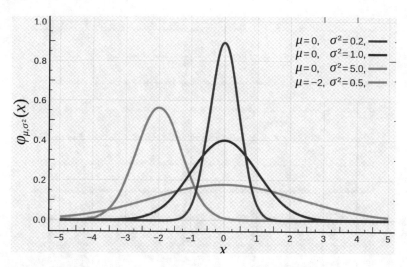

Four Gaussian distributions

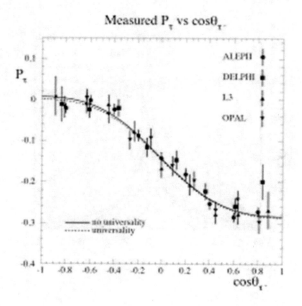

Cosinal distribution

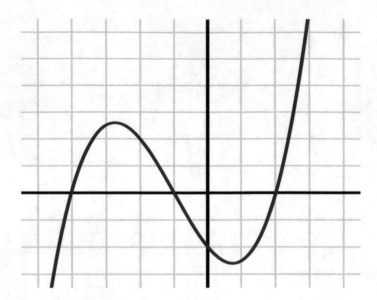

3rd order polynomial distribution

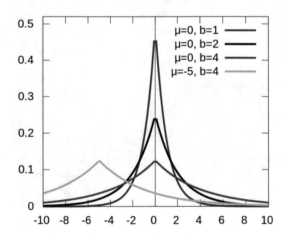

Multiple Laplace distributions

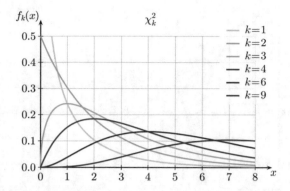

Additive-chi2 and Chi2 distributions

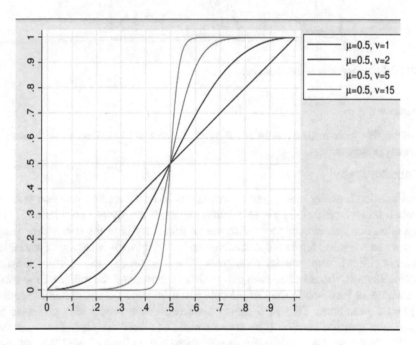

Sigmoid distributions

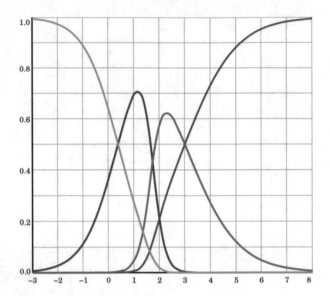

RBF (radial basis function) distributions

Misnomer

A name that is incorrectly applied. A phenomenon not uncommon with machine learning practices.

Multicollinearity

With multicollinearity one or more x-variables have a strong correlation with one another. It can be detected by a correlation matrix. If one by one correlations have a regression coefficient over 85%, then multicollinearity is in the data. Also a sharp increase in a t-value for the coefficient of an x-variable when another x-variable is removed from the model suggests the presence of multicollinearity. Multicollinearity does not reduce the predictive power or reliability of the model as a whole, at least within the sample data set; it only affects calculations regarding individual predictors. That is, a multivariate regression model with collinear predictors can indicate how well the entire bundle of predictors predicts the outcome variable, but it may not give valid results about any individual predictor, or about which predictors are redundant with respect to others.

How does ridge regression deal with multicollinearity? Ridge Regression is a technique for analyzing multiple regression data that suffer from multicollinearity. **By adding a degree of bias to the regression estimates**, ridge regression reduces the standard errors. It is hoped that the net effect will be to give estimates that are more reliable.

Multivariate Kernel Ridge Regression

With multivariate regression, also referred to as multi-task learning in machine learning, the goal is to recover a vector-valued function based on noisy observations. Although extensively studied and sometimes useful for the statistical analysis of data with multiple outcome variables, a theoretical study on multivariate nonlinear regressions is essentially missing to date, and it can therefore not be applied today in the nonlinear methodology of kernel ridge regression analyses.

O
Overfitting

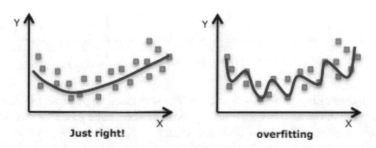

Overfitting will be observed, if data are wider than compatible with random sampling. Also if the modeling error is too close to a limited set of data points.

With kernel regressions y-values tendd to be overfitted, because they are added, and in this way continuous variables are discretized into discrete ones. A problem is that the results are only useful to the initial data but not to future datasets.

P
Problem with R Square Values

The problem with R Square values is that its magnitude is not an accurate and reliable predictor in datasets with small samples, and that additional analysis of variance tests are required for proving that sample sizes are adequate. For example, and old treatment is a possible predictor of a new treatment. It is tested with traditional linear regression. The underneath graph is a linear regression of the predictor values on the x-axis and the outcome values on the y-axis.

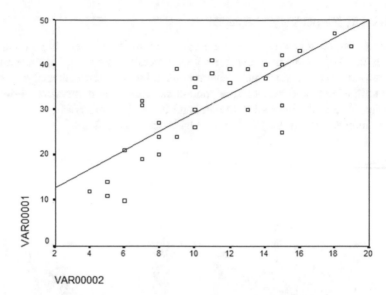

VAR00002

The underneath tables show that an traditional linear regression produces an R Square value of 0,630, meaning a strong correlation between predictor and outcome with 63% certainty about the outcome values if you know the predictor values. However, the strength of prediction is not only dependent on the magnitude of the R Square value, but also on the sample size of the data. With a small sample like 3 values exactly on the regression line the R Square value would be 100%. However, little certainty is provided, because the next few values may very well be far from the regression line. However, in the above graph we have 35 values. This gives much more certainty than does three values.

Analysis of variance (Anova) is helpful to the certainty about the outcome from the predictor (s) using an F (Fisher) test and a p-value <0,05, meaning that the "null hypothesis that the above 63% is not significantly different from zero %" is less than 5% and can thus be rejected. The p-value is even much less than 0,05, namely 0,000. See the underneath ANOVA table.

Model Summary

Model	R	R Square	Adjusted R Square	Std. Error of the Estimate
1	.794[a]	.630	.618	6.1590

a. Predictors: (Constant), VAR00002

ANOVA[b]

Model		Sum of Squares	df	Mean Square	F	Sig.
1	Regression	2128.393	1	2128.393	56.110	.000[a]
	Residual	1251.779	33	37.933		
	Total	3380.171	34			

a. Predictors: (Constant), VAR00002

b. Dependent Variable: VAR00001

Coefficients[a]

Model		Unstandardized Coefficients		Standardized Coefficients	t	Sig.
		B	Std. Error	Beta		
1	(Constant)	8.647	3.132		2.761	.009
	VAR00002	2.065	.276	.794	7.491	.000

a. Dependent Variable: VAR00001

A problem with kenrel ridge regressions is that ANOVAS are essentially missing.

Fortunately, however, kernel ridge regressions of small samples are typically rare, particularly with current big data machine learning datasets. With traditional regressions usually F statistics and p-values are used for rejecting the nullhypothesis of no difference from a zero predictive property. With kernel ridge regressions only R Squares can be used for the purpose.

Pseudo R Square Values

In order to compare the binary outcome logistic model versus the kernel ridge model, it would be convenient to compare the R Square values of either of the two methods. However, the logistic models do not provide the maths to produce R Square values. Instead, however, Cox and Snell (1989) proposed as alternative a pseudo R Square based on likelihood statistics. Its upperbound is like that of the true R Square 1000. However, It sometimes underestimates the certainty proportion in the binary data as given. In our data example the Cox and Snell pseudo R Square equals 0,700.

R
R Square values that, with a kernel ridge regression, can be negative

With traditional ordinary least square regression analysis the R Square values are between 0 and 1. How is it possible that with kernel ridge regression R square values can, obviously, be negative!!! The answer is, that with kernel ridge regression a negative R square value is possible for models where a datafit is worse than horizontal.

Ridge Regularization

A problem with kernel regressions is the problem of overfitted data, data wider than compatible with mathematical density distributions. It is corrected by ridge regularization which reduces overfitted b-values (regression coefficients) according to the equation $b_{ridge} = b / (1 + \lambda)$ with $\lambda =$ the shrinking factor.

S
Scikit-learn or Sklearn

The most useful library for machine learning in Python. I contains lots of useful tools for statistical modeling including classification, regression, clustering, dimensionality reduction. SPSS statistics version 28 0.1.0 used in the current textbook applies the Scikit-learn library in its menus for kernel ridge regression analysis.

Study Stream

A study stream is a dedicated place online where students can study with one another, and where they can share each others' tips and more.

Stream Study

Idem Study Stream.

T
Test-Retest Reliability

Test-retest reliability is a measure of reliability obtained by administering the same test twice over a period of time to a group of individuals. The scores from Time 1 and Time 2 can then be correlated in order to evaluate the test for stability over time. Does test-retest measure reliability or validity? Having good test re-test reliability signifies the internal validity of a test and ensures that the measurements obtained in one sitting are both representative and stable over time.

Note Many more new terminologies are applied in the current edition, but most of them have been explained directly in the text.

5.3   References

All of the chapters of the current edition start with a brief review of the traditional analytic method of the different regression methods prior to the review of the relevant kernel ridge regression method. For the purpose, generally, data examples are used from the recent edition "Regression Analyses in Clinical Research for Starters and 2nd Levelers 2nd Edition, Springer Heidelberg Germany 2021", by the same authors. For a better understanding of differences between traditional and kernel regressions, readers may benefit from the study of this edition first.

To readers requesting still more background, theoretical and mathematical information of computations given, several textbooks complementary to the current production and written by the same authors are available: Statistics applied to clinical studies 5th edition, 2012, Machine learning in medicine a complete overview 2nd edition, 2020, SPSS for starters and 2nd levelers 2nd edition, 2015, Clinical data analysis on a pocket calculator 2nd edition, 2016, Understanding clinical data analysis from published research, 2016, all of them edited by Springer Heidelberg Germany.

Chapter 6
Effect on Being Blind of Age/Sex Adjusted Mortality of Onchocerciasis Patients in 12,816 Personyears, Traditional vs Kernel Ridge Regression

Abstract In 12,816 onchocerciasis patientyears it was studied whether age, sex and nonblind mortality are predictors of blind deaths. The overall R Square value is 0,994, which means 94,4% certainty about the prediction of the outcome by the above three predictors. The presence of multicollinearity in the data was suspected, and confirmed as assessed with one-by-one linear regressions. Kernel ridge regression is a technique for analyzing multiple regression data that suffer from multicollinearity. By adding a degree of bias to the regression estimates, ridge regression reduces the standard errors, giving rise to a more accurate without loss of sensitivity of testing. The best fit kernel density model was obtained by the polynomial kernel ridge regression with an R Square value of no less than 1,00 (100% certainty of prediction), which is better fitted than the one from the above flawed traditional linear regression of 0,994. And so, kernel ridge regression provided not only a less flawed analysis, but also provided a better R-Square value of prediction of 1,00 instead of 0,994.

Keywords Kernel ridge regression · R Square values · Multicollinearity · Polynomial kernel density model

6.1 Summary

In observational research event rates are often very much age and sex dependent and a model routinely adjusting these confounders is welcome. A data example will be given from Kirkwood and Sterne (Standardization, in: Medical Statistics, Chap. 25, Blackwell Science, Oxford UK 2003). The authors studied in 12,816 onchocerciasis patientyears, whether age, sex and nonblind mortality are predictors of blind deaths.

Supplementary Information The online version contains supplementary material available at [https://doi.org/10.1007/978-3-031-10717-7_6].

6.1.1   Summaries of Traditional Regressions

The overall R Square value is 0,994, which means 94,4% certainty about the prediction of the outcome by the three predictors. This is significantly different from zero % at p = 0,000. However, on clinical grounds the presence of multicollinearity in the data was suspected. And, therefore, its presence was assessed with one-by-one linear regressions. If CC (correlation coefficient) > 0,85, then multicollinearity is in the data, and the model is no longer entirely valid. One of the two variables responsible must be removed. For assessment the underneath commands are required.

The correlation coefficient (CC) between gender and ageclass is 1000, much more than 0,85. Also five more one-by-one correlations were over 0,85. The traditional regression model is no longer valid. It model must be replaced with a ridge regression. Ridge regression is a technique for analyzing multiple regression data that suffer from multicollinearity. By adding a degree of bias to the regression estimates, ridge regression reduces the standard errors. It is hoped that the net effect will be to give estimates that are more accurate without loss of sensitivity of testing.

6.1.2   Summaries of Kernel Ridge Regressions

The best fit kernel density model was obtained by the polynomial kernel ridge regression with an R Square value of no less than 1,00 (100% certainty of prediction), which is better fitted than the one from the above flawed traditional linear regression of 0,994. And so, kernel ridge regression provided not only a less flawed analysis, but also provided a better R-Square value of prediction of 1,00 instead of 0,994.

We note, that several R Squares of the different kernel density models were negative. These values are from ill fitting kernel density models. The phenomenon of negative kernel density R Squares is explained in the Chap. 5 entitled "Some Terminology".

6.2   Introduction

In observational research event rates are often very much age and sex dependent and a model routinely adjusting these confounders is welcome. A data example will be given from Kirkwood and Sterne (Standardization, in: Medical Statistics, Chap. 25, Blackwell Science, Oxford UK 2003). The authors studied in 12,816 personyears, whether age, sex and nonblind mortality are predictors of blind deaths in onchocerciasis patients.

6.3 Data Example

Variables			
1	2	3	4
gender	ageclass	nonblinddeaths personyears	blinddeaths personyears
1,00	1,00	7,90	25,00
1,00	2,00	13,20	40,90
1,00	3,00	17,90	53,30
1,00	4,00	32,80	101,30
,00	1,00	7,40	23,80
,00	2,00	13,70	43,50
,00	3,00	17,20	47,60
,00	4,00	29,60	96,80

The datafile is in SpringerLink supplementary files, and is entitled "blind".

It should be downloaded in your computer installed with SPSS statistical software 2022 version 28 0.1.0.

Variables 1-4:

1. gender = 1 male, 0 female
2. ageclass = 1: 30-39, 2: 40-49, 3:50-59, 4: 60 and over.
3. nonblinddeathspersonyears = numbers deaths per 1000 personyears per year in the nonblinds
4. blinddeathspersonyears = numbers deaths per 1000 personyears per year in the blinds

6.4 Traditional Linear Regression

Traditional linear regression with blind deaths as outcome and nonblinddeaths, gender and age as predictors is performed.

Command:

Analyze....regression....Linear....Dependent: blinddeathsperssonyears....Independent(s): nonblinddeathspersonyears, gender, ageclass....click OK.

The underneath tables come up.

Model Summary

Model	R	R Square	Adjusted R Square	Std. Error of the Estimate
1	,997[a]	,994	,990	2,96764

a. Predictors: (Constant), nonblinddeathspersonyears, gender, ageclass

ANOVA[b]

Model		Sum of Squares	df	Mean Square	F	Sig.
1	Regression	6110,247	3	2036,749	231,268	,000[a]
	Residual	35,228	4	8,807		
	Total	6145,475	7			

a. Predictors: (Constant), nonblinddeathspersonyears, gender, ageclass
b. Dependent Variable: blinddeathspersonyears

Coefficients[a]

Model		Unstandardized Coefficients		Standardized Coefficients	t	Sig.
		B	Std. Error	Beta		
1	(Constant)	1,513	2,873		,527	,626
	gender	-1,579	2,140	-,028	-,738	,502
	ageclass	-5,754	3,363	-,232	-1,711	,162
	nonblinddeathspersonyears	3,876	,432	1,219	8,972	,001

a. Dependent Variable: blinddeathspersonyears

The overall R Square value is 0,994, which means 94,4% certainty about the prediction of the outcome by the three predictors. This is significantly different from zero % at p = 0,000. However, on clinical grounds the presence of multicollinearity in the data was suspected. And, therefore, its presence was assessed with one-by-one linear regressions. If CC > 0,85, then multicollinearity is in the data, and the model is no longer entirely valid. One of the two variables responsible must be removed. For assessment the underneath commands are required.

Command:

Analyze....Correlate....Bivariate....Variables: enter blinddeaths, gender, ageclass, nonblinddeaths....mark Pearson....click OK.

A one-by-one linear regression diagram with Pearson correlation coefficients is in the output.

Correlations

		gender	ageclass	nonblinddeat hspersonyear s	blinddeathsp ersonyears
gender	Pearson Correlation	1	,000	,056	,040
	Sig. (2-tailed)		1,000	,895	,926
	N	8	8	8	8
ageclass	Pearson Correlation	,000	1	,959[**]	,937[**]
	Sig. (2-tailed)	1,000		,000	,001
	N	8	8	8	8
nonblinddeathspersonye ars	Pearson Correlation	,056	,959[**]	1	,995[**]
	Sig. (2-tailed)	,895	,000		,000
	N	8	8	8	8
blinddeathspersonyears	Pearson Correlation	,040	,937[**]	,995[**]	1
	Sig. (2-tailed)	,926	,001	,000	
	N	8	8	8	8

[**]. Correlation is significant at the 0.01 level (2-tailed).

The correlation coefficient between gender and ageclass is 1000, much more than 0,85. Also five more one-by-one correlations were over 0,85.

The model is no longer valid. The traditional analysis model must be replaced with a ridge regression. Ridge Regression is a technique for analyzing multiple regression data that suffer from multicollinearity. By adding a degree of bias to the regression estimates, ridge regression reduces the standard errors. It is hoped that the net effect will be to give estimates that are more accurate without loss of sensitivity of testing.

6.5 Kernel Ridge Regressions

Command:

Analyze....Regression....Kernel Ridge Regression....Dependent: blinddeathspersonyears....Independent(s): nonblinddeathspersonyears, gender, ageclass....click Linear....click OK.

Model Summary[a,b]

Kernel	Alpha	R Square
Linear	1,000	,993

a. Dependent Variable:
   blinddeathspersonyears

b. Model: gender,
   ageclass,
   nonblinddeathspersonye
   ars

The above R Square was similar to the one with traditional linear regression, 0,993 versus 0,994.

More kernel density models are assessed. Similar commands are given.

Model Summary[a,b]

Kernel	Alpha	R Square
Additive_chi2	1,000	,983

a. Dependent Variable:
   blinddeathspersonyears

b. Model: gender, ageclass,
   nonblinddeathspersonyears

Model Summary[a,b]

Kernel	Alpha	Gamma	R Square
Chi2	1,000	1,000	,232

a. Dependent Variable:
   blinddeathspersonyears

b. Model: gender, ageclass,
   nonblinddeathspersonyears

Model Summary[a,b]

Kernel	Alpha	R Square
Cosine	1,000	-,041

a. Dependent Variable:
blinddeathspersonyears

b. Model: gender, ageclass,
nonblinddeathspersonye
ars

Model Summary[a,b]

Kernel	Alpha	Gamma	R Square
Laplacian	1,000	,333	,187

a. Dependent Variable:
blinddeathspersonyears

b. Model: gender, ageclass,
nonblinddeathspersonyears

Model Summary[a,b]

Kernel	Alpha	Gamma	R Square
RBF	1,000	,333	-,012

a. Dependent Variable:
blinddeathspersonyears

b. Model: gender, ageclass,
nonblinddeathspersonyears

Model Summary[a,b]

Kernel	Alpha	Gamma	Coef0	Degree	R Square
Polynomial	1,000	,333	1,000	3,000	1,000

a. Dependent Variable: blinddeathspersonyears

b. Model: gender, ageclass, nonblinddeathspersonyears

Model Summary[a,b]

Kernel	Alpha	Gamma	R Square
RBF	1,000	,333	-,012

a. Dependent Variable: blinddeathspersonyears

b. Model: gender, ageclass, nonblinddeathspersonyears

Model Summary[a,b]

Kernel	Alpha	Gamma	Coef0	R Square
Sigmoid	1,000	,333	1,000	-,047

a. Dependent Variable: blinddeathspersonyears

b. Model: gender, ageclass, nonblinddeathspersonyears

Obviously, the best fit kernel density model was obtained by the polynomial kernel ridge regression with an R Square value of no less than 1,00 (100% certainty of prediction), which is better fitted than the one from the above flawed traditional linear regression of 0,994. And so, kernel ridge regression provided not only a less flawed analysis, but also provided a better R-Square value of prediction of 1,00 instead of 0,994.

Note that several R Squares of the different kernel density models were negative. These values are from ill fitting kernel density models. The phenomenon of negative kernel density R Squares is explained in the Chap. 5 entitled "Some Terminology".

We should add that the above analyses use eight different kernel density models. Graphical presentations of all of them are in the Chap. 5 entitled "Some Terminologies".

6.6 Conclusion

6.6.1 Summaries of Traditional Regressions

The overall R Square value is 0,994, which means 94,4% certainty about the prediction of the outcome by the three predictors. This is significantly different from zero % at $p = 0,000$. However, on clinical grounds the presence of multicollinearity in the data was suspected. And, therefore, its presence was assessed with one-by-one linear regressions. If CC > 0,85, then multicollinearity is in the data, and the model is no longer entirely valid. One of the two variables responsible must be removed. For assessment the underneath commands are required.

The correlation coefficient between gender and ageclass is 1000, much more than. 0,85. Also five more one-by-one correlations were over 0,85. The model is no longer valid. The traditional analysis model must be replaced with a ridge regression. Ridge Regression is a technique for analyzing multiple regression data that suffer from multicollinearity. By adding a degree of bias to the regression estimates, ridge regression reduces the standard errors. It is hoped that the net effect will be to give estimates that are more accurate without loss of sensitivity of testing.

6.6.2 Summaries of Kernel Ridge Regressions

The best fit kernel density model was obtained by the polynomial kernel ridge regression with an R Square value of no less than 1,00 (100% certainty of prediction), which is better fitted than the one from the above flawed traditional linear regression of 0,994. And so, kernel ridge regression provided not only a less flawed analysis, but also provided a better R-Square value of prediction of 1,00 instead of 0,994.

We note, that several R Squares of the different kernel density models were negative. These values are from ill fitting kernel density models. The phenomenon of negative kernel density R Squares is explained in the Chap. 5 entitled "Some Terminology".

6.7 References

All of the chapters of the current edition start with a brief review of the traditional analytic method of the different regression methods prior to the review of the relevant kernel ridge regression method. For the purpose, generally, data examples are used from the recent edition "Regression Analyses in Clinical Research for Starters and 2nd Levelers 2nd Edition, Springer Heidelberg Germany 2021", by the same authors. For a better understanding of differences between traditional and kernel ridge regressions, readers may benefit from the study of this edition first.

To readers requesting still more background, theoretical and mathematical information of computations given, several textbooks complementary to the current production and written by the same authors are available: Statistics applied to clinical studies 5th edition, 2012, Machine learning in medicine a complete overview 2nd edition, 2020, SPSS for starters and 2nd levelers 2nd edition, 2015, Clinical data analysis on a pocket calculator 2nd edition, 2016, Understanding clinical data analysis from published research, 2016, all of them edited by Springer Heidelberg Germany.

Chapter 7
Effect of Old Treatment on New Treatment, 35 Patients, Traditional Regressions vs Kernel Ridge Regressions

Abstract In this chapter the numbers of stools on a new laxative as outcome and the numbers of stools on the old laxative as predictor in 35 constipated patients was used as data example for testing the effects of simple linear regression, and quantile regression against kernel ridge regression. With traditional linear regression the old treatment was not a strong predictor of the new treatment with an overall R Square value of 0,219 (21,9% certainty about the outcome knowing the predictor. R square values under 25% are considered to be weak predictors. The best fit pseudo R Square values with the quantile regressions 0,1, 0,2, 0,3 were respectively 0,310, 0.259, and 0,220, all of them slightly "betterfit" than the traditional regression R Square of 0,219. Kernel ridge regression provided an R Square value of no less than 0,831 (83,1% certainty is predicted by the predictors about the outcome).

Keywords Simple linear regression · Kernel ridge regression · Quantile regression · R square values · Pseudo R Square values

7.1 Summary

The history, background, and the development of the analytical data models of traditional and kernel regressions have already been addressed in the Chaps. 1, 2 and 3. In this chapter the numbers of stools on a new laxative as outcome and the numbers of stools on the old laxative as predictor in 35 constipated patients will be used as data example. Simple linear regression produced a borderline p-value of 0.049, not a very powerful result. More statistical power was desirable. A GENLIN (Generalized linear regression-generalized linear regression) procedure can be followed using maximum likelihood estimators and/or robust regression.

Supplementary Information The online version contains supplementary material available at [https://doi.org/10.1007/978-3-031-10717-7_7].

7.1.1   Summaries of Traditional Regressions

With traditional linear regression the old treatment was a borderline significant predictor at p = 0.048 of the novel treatment. More statistical power was desirable.

Neither was the overall R Square value of 0,219 a strong predictor (21,9% certainty about the outcome knowing the predictors. R square values under 25% are considered to be weak predictors. A problem with robust regressions in GENLIN is that no R Square values are provided, and, so, the strength cannot be precisely tested against kernel ridge models. Quantile regression does not provide R Square values, but pseudo R Square values can do the job (see the Chap. 5 entitled "Some Terminology"). And, so, its datafit can somewhat better be compared against kernel ridge regression and traditional linear regression. The best fit pseudo R Square values were with the quantiles 0,1, 0,2, 0,3,and the respective values 0,310, 0.259, and 0,220. All of them were slightly "betterfit" than the traditional regression R Square of 0,219. Yet all them were pretty poor predictors giving only 31,1%, 25,9%, and 22,0% certainty about the outcome knowing the predictors, and stronger prediction models are welcome. Kernel ridge regressions will be applied for the purpose next.

7.1.2   Summaries of Kernel Ridge Regressions

Obviously, the polynomial kernel density provided the best fit R Square value of no less than 0,831 (83,1% certainty is predicted by the predictors about the outcome. The scatterplot of the predicted newtreat values versus predicted ones shows a very nice linear pattern of the polynomial kernel ridge regression, while the spread of the residuals had a constant pattern.

7.2   Introduction

The history, background, and the development of the analytical data models have already been addressed in the Chaps. 1, 2 and 3. In this Chap. the numbers of stools on a new laxative as outcome and the numbers of stools on the old laxative as predictor in 35 constipation patients will be used as data example. Simple linear regression produced a borderline p-value of 0.049, with an R Square value of 0,219, i.e., 21,9% certainty about the outcome knowing the predictor variables, not a very powerful result. We should add, that the patients no. 13 and 27 had outlier "oldtreat" scores, and the analists here suggested the possibility of typing errors. But this was not confirmed, because of, otherwise, normal explanations for the observations.

7.3   Data Example

35 pts

newtreat	oldtreat	agecats	patientnumber
24.00	8.00	2.00	1.00
30.00	13.00	2.00	2.00
25.00	15.00	2.00	3.00
35.00	10.00	3.00	4.00
39.00	9.00	3.00	5.00
30.00	10.00	3.00	6.00
27.00	8.00	1.00	7.00
14.00	5..00	1.00	8.00
39.00	13.00	1.00	9.00
42.00	15.00	1.00	10.00
41.00	11.00	1.00	11.00
38.00	11.00	2.00	12.00
39.00	112.00	2.00	13.00
37.00	10.00	3.00	14.00
47.00	18.00	3.00	15.00
30.00	13.00	2.00	16.00
36.00	12.00	2.00	17.00
12.00	4.00	2.00	18.00
26.00	10.00	2.00	19.00
20.00	8.00	1.00	20.00
43.00	16.00	3.00	21.00
31.00	15.00	2.00	22.00
40.00	114.00	2.00	23.00
31.00	7.00	2.00	24.00
36.00	12.00	3.00	25.00
21.00	6.00	2.00	26.00
44.00	19.00	3.00	27.00
11.00	5.00	2.00	28.00
27.00	8.00	2.00	29.00
24.00	9.00	2.00	30.00
40.00	15.00	1.00	31.00
32.00	7.00	2.00	32.00
10.00	6.00	2.00	33.00
37.00	14.00	3.00	34.00
19.00	7.00	2.00	35.00

newtreat = new treatment
oldtreat = old treatment
agecats = age categories
patientnumber = patient number

First, a traditional linear regression analysis will be performed.

7.4   Traditional Linear Regression

The datafile is in SpringerLink supplementary files, and it is entitled "robustregression". Start by downloading the datafile into your computer installed with SPSS statistical software 2022 version 28 0.1.0.

Command:

Analyze....Menu....Regression....Linear....Dependent: enter newtreat.... Independent (s): oldtreat, age, patientid....click OK.
   In the output are the underneath tables.

Model Summary

Model	R	R Square	Adjusted R Square	Std. Error of the Estimate
1	,468[a]	,219	,144	9,22761

a. Predictors: (Constant), patient id, old treatment, age

ANOVA[b]

Model		Sum of Squares	df	Mean Square	F	Sig.
1	Regression	740,561	3	246,854	2,899	,051[a]
	Residual	2639,611	31	85,149		
	Total	3380,171	34			

a. Predictors: (Constant), patient id, old treatment, age
b. Dependent Variable: new treatment

Coefficients[a]

Model		Unstandardized Coefficients		Standardized Coefficients	t	Sig.
		B	Std. Error	Beta		
1	(Constant)	23,364	5,653		4,133	,000
	old treatment	,134	,065	,327	2,062	,048
	age	4,127	2,323	,283	1,777	,085
	patient id	-,182	,155	-,187	-1,176	,249

a. Dependent Variable: new treatment

With traditional linear regression the old treatment was a borderline significant predictor at p = 0,048 of the novel treatment. More statistical power was desirable.

Neither was the overall R Square value of 0,219 a strong predictor (21,9% certainty about the outcome knowing the predictors. R square values under 25% are considered to be weak predictors.

7.5   Robust Regression

Instead of traditional linear regression, a GENLIN (Generalized linear regression-generalized linear regression) procedure can be followed in SPSS using maximum likelihood estimators instead of traditional F- and t-tests. The results are likely to produce a bit better precision.

Command:

Generalized Linear Models....Generalized Linear Models....mark: Custom.... Distribution: select Normal....Link function: select identity....Response: Dependent Variable: enter new treatment....Predictors: Factors: enter old treatment....Model: Model: enter oldtreat....Estimation: mark Model-based Estimator....click OK.

The underneath table is in the output.

Parameter Estimates

Parameter	B	Std. Error	95% Wald Confidence Interval		Hypothesis Test		
			Lower	Upper	Wald Chi-Square	df	Sig.
(Intercept)	40,000	4,2650	31,641	48,359	87,958	1	,000
[oldtreat=4,00]	-28,000	6,0317	-39,822	-16,178	21,550	1	,000
[oldtreat=5,00]	-27,500	5,2236	-37,738	-17,262	27,716	1	,000
[oldtreat=6,00]	-24,500	5,2236	-34,738	-14,262	21,999	1	,000
[oldtreat=7,00]	-12,667	4,9248	-22,319	-3,014	6,615	1	,010
[oldtreat=8,00]	-15,500	4,7684	-24,846	-6,154	10,566	1	,001
[oldtreat=9,00]	-8,500	5,2236	-18,738	1,738	2,648	1	,104
[oldtreat=10,00]	-8,000	4,7684	-17,346	1,346	2,815	1	,093
[oldtreat=11,00]	-,500	5,2236	-10,738	9,738	,009	1	,924
[oldtreat=12,00]	-4,000	5,2236	-14,238	6,238	,586	1	,444
[oldtreat=13,00]	-7,000	4,9248	-16,652	2,652	2,020	1	,155
[oldtreat=14,00]	-3,000	6,0317	-14,822	8,822	,247	1	,619
[oldtreat=15,00]	-5,500	4,7684	-14,846	3,846	1,330	1	,249
[oldtreat=16,00]	3,000	6,0317	-8,822	14,822	,247	1	,619
[oldtreat=18,00]	7,000	6,0317	-4,822	18,822	1,347	1	,246
[oldtreat=19,00]	4,000	6,0317	-7,822	15,822	,440	1	,507
[oldtreat=112,00]	-1,000	6,0317	-12,822	10,822	,027	1	,868
[oldtreat=114,00]	0[a]	.	.	.	.	.	.
(Scale)	18,190[b]	4,3484	11,386	29,062			

Dependent Variable: new treatment
Model: (Intercept), oldtreat

a. Set to zero because this parameter is redundant.

b. Maximum likelihood estimate.

Even better precision may be obtained by the use of **robust** standard errors, called the Hubert-White estimators by SPSS statistical software.

Command:

Generalized Linear Models....Generalized Linear Models....mark: Custom.... Distribution: select Normal....Link function: select identity.... Response: Dependent Variable: enter new treatment....Predictors: Factors: enter old treatment....Model: Model: enter newtreat....Estimation: mark Robust Estimator....click OK.

The underneath table is in the output.

Parameter Estimates

Parameter	B	Std. Error	95% Wald Confidence Interval		Hypothesis Test		
			Lower	Upper	Wald Chi-Square	df	Sig.
(Intercept)	40,000	.	.	.	.	1	,000
[oldtreat=4,00]	-28,000	.	.	.	.	1	,000
[oldtreat=5,00]	-27,500	1,0607	-29,579	-25,421	672,222	1	,000
[oldtreat=6,00]	-24,500	3,8891	-32,122	-16,878	39,686	1	,000
[oldtreat=7,00]	-12,667	3,4102	-19,351	-5,983	13,796	1	,000
[oldtreat=8,00]	-15,500	1,4361	-18,315	-12,685	116,485	1	,000
[oldtreat=9,00]	-8,500	5,3033	-18,894	1,894	2,569	1	,109
[oldtreat=10,00]	-8,000	2,1506	-12,215	-3,785	13,838	1	,000
[oldtreat=11,00]	-,500	1,0607	-2,579	1,579	,222	1	,637
[oldtreat=12,00]	-4,000	.	.	.	.	1	,000
[oldtreat=13,00]	-7,000	2,4495	-11,801	-2,199	8,167	1	,004
[oldtreat=14,00]	-3,000	.	.	.	.	1	,000
[oldtreat=15,00]	-5,500	3,4369	-12,236	1,236	2,561	1	,110
[oldtreat=16,00]	3,000	.	.	.	.	1	,000
[oldtreat=18,00]	7,000	.	.	.	.	1	,000
[oldtreat=19,00]	4,000	.	.	.	.	1	,000
[oldtreat=112,00]	-1,000	.	.	.	.	1	,000
[oldtreat=114,00]	0[a]	.	.	.	.	.	.
(Scale)	18,190[b]	4,3484	11,386	29,062			

Dependent Variable: new treatment
Model: (Intercept), oldtreat

a. Set to zero because this parameter is redundant.

b. Maximum likelihood estimate.

Out of the stool scores with the old treatment, 6 scores provided p-values of <0,05 with the Model-based Estimator, while it produced up to 14 p-values <0,05 with the Robust Estimator. If your results are borderline significant, like in the above example, then loglikelihood regression testing and robust regression testing can, obviously, provide better statistics, and, thus, better statistical power of testing, than traditional testing can. These highly significant results from this pretty small data sample, are they credible? Maybe, not entirely, but robust regressions is a sophisticated model, where a continuous predictor variable is restructured into multiple binary variables. If we can believe in the robust methodology, then we will have to accept the result. Nonetheless, the simple univariate linear model has been replaced by the software with a multiple variables linear model. Regression analysis with multiple predictors has long been interpreted as a rather inferior type of statistical data analysis due to multiple testing and multiple type I errors. However, much has changed since regressions have entered the field of causality research, like, for example, with structural equation modeling. The EMA (European Medicines Agency) has approved the use of a few predictive variables in confirmative controlled clinical trials. And although many data analists are still doubtful, the time has come that multiple variables regressions have received at least some serious attention. The p-values may not have the heavy interpretation that prospective blinded trial p-values have, but multiple variables regressions are no longer adjusted for multiple testing, because they are assumed to stem from a family of null hypotheses with many interactions within a single experiment. A problem with robust regressions in GENLIN is that no R Square values are provided, and, so, the strength cannot be precisely tested against kernel ridge models. As an alternative quantile regression is possible for comparison against kernel ridge regression.

7.6   Quantile Regressions

The methodology is pretty new, and a textbook for clinicians and health workers entitled "Quantile regression in clinical research" has been published by Springer Heidelberg Germany in 2021 (edited by the authors of the current work). A brief introduction of the method is underneath.

Quantile regression needs not adjusted for multiple testing. The quantile regression analysis often includes around ten null-hypothesis tests or so, with quantiles ranging from 0,1 to 0,9. However, again a single question is answered: which one of them gives the best result. Better statistics may be obtained with the help of quantile regression, and, in addition, a better insight in the relationships between predictor and outcome variables.

Quantiles (fraction of the data) and percentiles (percentage of the data) are identical terms. Usually, linear regression assumes, that the y-variable has a normal distribution, and can be summarized by means. The least square computation of the regression coefficient is obtained by the use of the means of the x- and the y-values.

Y_m = mean of observed Y-values.
X_m = mean of observed X-values.

$$B = \frac{\Sigma(X - X_m)(Y - Y_m)}{\Sigma(X - X_m)^2} = \frac{\Sigma XY - n\,X_m Y_m}{\Sigma X^2 - n\,X_m{}^2}$$

If a normal distribution is not assumed, summaries of quantiles, like the 0,10, 0,20, 0,30 quantile etc. will be an adequate alternative for computing regression coefficients. The data file as used in the above section, is used again. The mean x and y values are replaced with either median x and y values or 0,1 to 0,9 quantile x and y values.

Start by entering the datafile in your computer mounted with SPSS statistical software 2022 version 28.0.1.0.

Command:

Analyze....Regression....Quantile Regression....click Target Variable: enter new treatment....click Covariate(s): enter old treatment, age, patientno....click Criteria.... mark Specify single quantiles....Quantile value(s): 0,1 Add, 0,2 Add, 0,3 Add, 0,4 Add, 0,5 Add, 0,6 Add, 0,7 Add, 0,8 Add, 0,9 Add....click Continue...click Display....Print mark parameter estimates....mark Plot or tabulate top 3 effects....in Model Effects move old treat to Prediction Lines....click Continue....click OK.

In the output sheets are the underneath interactive tables and graphs.

Model Quality[a,b,c]

	q=0,1	q=0,2	q=0,3	q=0,4	q=0,5	q=0,6	q=0,7	q=0,8	q=0,9
Pseudo R Squared	,310	,259	,220	,195	,174	,130	,047	,014	,051
Mean Absolute Error (MAE)	12,2135	8,9997	7,7200	6,8982	6,6990	6,8551	8,1722	9,1217	12,3435

a. Dependent Variable: new treatment

b. Model: (Intercept), old treatment, age, patient id

c. Method: Simplex algorithm

Parameter Estimates by Different Quantiles[a,b]

Parameter	q=0,1	q=0,2	q=0,3	q=0,4	q=0,5	q=0,6	q=0,7	q=0,8	q=0,9
(Intercept)	7,398	5,814	3,444	12,274	16,515	21,631	32,350	39,484	38,202
old treatment	,193	,161	,141	,121	,099	,084	,044	,021	,226
age	7,740	8,522	10,089	7,200	6,099	5,132	2,213	-,103	1,357
patient id	-,425	-,142	-,030	-,022	9,598E-17	-,062	-,077	-,073	-,095

a. Dependent Variable: new treatment

b. Model: (Intercept), old treatment, age, patient id

Quantile = 0,1

Parameter Estimates[a,b]

Parameter	Coefficient	Std. Error	t	df	Sig.	95% Confidence Interval	
						Lower Bound	Upper Bound
(Intercept)	7,398	4,4799	1,651	31	,109	-1,739	16,535
old treatment	,193	,0514	3,763	31	<,001	,089	,298
age	7,740	1,8409	4,205	31	<,001	3,986	11,495
patient id	-,425	,1228	-3,464	31	,002	-,676	-,175

a. Dependent Variable: new treatment

b. Model: (Intercept), old treatment, age, patient id

Quantile = 0,2

Parameter Estimates[a,b]

Parameter	Coefficient	Std. Error	t	df	Sig.	95% Confidence Interval	
						Lower Bound	Upper Bound
(Intercept)	5,814	9,3917	,619	31	,540	-13,341	24,968
old treatment	,161	,1077	1,492	31	,146	-,059	,380
age	8,522	3,8594	2,208	31	,035	,650	16,393
patient id	-,142	,2574	-,553	31	,584	-,667	,383

a. Dependent Variable: new treatment

b. Model: (Intercept), old treatment, age, patient id

Quantile = 0,3

Parameter Estimates[a,b]

Parameter	Coefficient	Std. Error	t	df	Sig.	95% Confidence Interval	
						Lower Bound	Upper Bound
(Intercept)	3,444	9,2871	,371	31	,713	-15,497	22,386
old treatment	,141	,1065	1,321	31	,196	-,076	,358
age	10,089	3,8164	2,644	31	,013	2,305	17,873
patient id	-,030	,2546	-,116	31	,908	-,549	,490

a. Dependent Variable: new treatment

b. Model: (Intercept), old treatment, age, patient id

Unlike the above robust regression, quantile regression provides pseudo R Square values based on loglikelihood ratios, and, so, its datafit can somewhat better be compared against kernel ridge regression and traditional linear regression. The best fit pseudo R Square values were with the quantiles 0,1, 0,2, 0,3,and the respective values 0,310, 0.259, and 0,220. All of them were slightly "betterfit" than the traditional regression R Square of 0,219. Yet all of them were yet pretty poor predictors giving only 31,1%, 25,9%, and 22,0% certainty about the outcome knowing the predictors, and stronger prediction models are welcome. Kernel ridge regressions will be applied for the purpose next.

7.7   Kernel Ridge Regression

Command:

Analyze....Regression....Kernel ridge regression....Dependent: new treatment.... Independent(s): old treatment, age, patient id....click Linear....click OK.

In the output is the table below.

Model Summary[a,b]

Kernel	Alpha	R Square
Linear	1,000	-,212

a. Dependent Variable: newtreat

b. Model: oldtreat, agecategories, patientno

The R square of the Linear kernel ridge regression density model was $-0,212$. This means a very poor datafit. We should stress that with kernel ridge regressions R Square may sometimes be negative. This is explained in the Chap. 5 entitled "Some Terminologies". Note: the linear kernel density model uses Gaussian distributions. See also the Chap. 5 "four Gaussian distributions".

More kernel density models are assessed giving similar commands.

Model Summary[a,b]

Kernel	Alpha	R Square
Additive_chi2	1,000	,763

a. Dependent Variable: newtreat

b. Model: oldtreat, agecategories, patientno

Model Summary[a,b]

Kernel	Alpha	Gamma	R Square
Chi2	1,000	1,000	,034

a. Dependent Variable: newtreat

b. Model: oldtreat, agecategories, patientno

Model Summary[a,b]

Kernel	Alpha	R Square
Cosine	1,000	,374

a. Dependent Variable: newtreat

b. Model: oldtreat, agecategories, patientno

Model Summary[a,b]

Kernel	Alpha	Gamma	R Square
Laplacian	1,000	,333	-,427

a. Dependent Variable: newtreat

b. Model: oldtreat, agecategories, patientno

Model Summary[a,b]

Kernel	Alpha	Gamma	Coef0	Degree	R Square
Polynomial	1,000	,333	1,000	3,000	,831

a. Dependent Variable: newtreat

b. Model: oldtreat, agecategories, patientno

Model Summary[a,b]

Kernel	Alpha	Gamma	R Square
RBF	1,000	,333	-1,167

a. Dependent Variable: newtreat

b. Model: oldtreat, agecategories, patientno

Model Summary[a,b]

Kernel	Alpha	Gamma	Coef0	R Square
Sigmoid	1,000	,333	1,000	-,008

a. Dependent Variable: newtreat

b. Model: oldtreat, agecategories, patientno

Obviously, the polynomial kernel density provided the best fit R Square value of no less than 0,831 (83,1% certainty is predicted by the predictors about the outcome.

The above eight kernel ridge analyses used eight different kernel density models, that are graphically displayed in the Chap. 5, entitled "Some Terminology".

We can also command the software to draw scatterplots of predicted outcome values and predicted residual values, and request numerical values.

Command:

Analyze....Regression....Kernel ridge regression....Dependent: new treatment.... Independent(s): old treatment, age, patient id....click Options mark Observed vs Predicted....Mark Predicted values....click Continue....click Linear....click OK..

In the output two graphs are given.

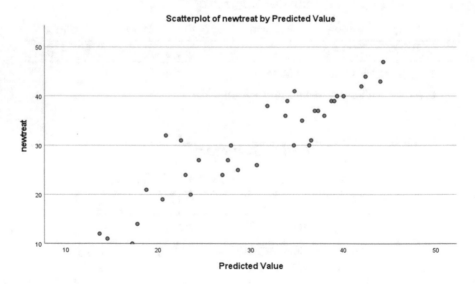

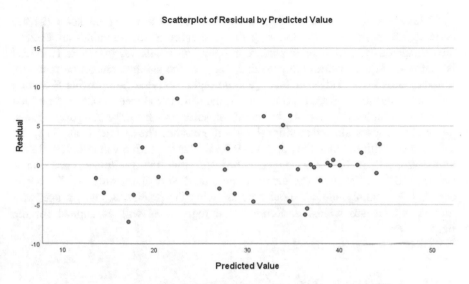

The scatterplot of the predicted newtreat values versus predicted ones shows a very nice linear pattern of the polynomial kernel ridge regression, while the spread of the residuals does not increase with increase of the predicted values.

Returning to the dataview screen, it is shown that a fifth data column has been added. It consists of predicted newtreat values.

newtreat	oldtreat	agecategories	patientno	predicted newtreat
24,00	8,00	2,00	1,00	22,9835
30,00	13,00	2,00	2,00	27,8731
25,00	15,00	2,00	3,00	28,6248
35,00	10,00	3,00	4,00	35,5105
39,00	9,00	3,00	5,00	33,8953
30,00	10,00	3,00	6,00	36,2904
27,00	8,00	1,00	7,00	27,5460
14,00	5,00	1,00	8,00	17,7692
39,00	13,00	1,00	9,00	38,7049
42,00	15,00	1,00	10,00	41,9290
41,00	11,00	1,00	11,00	34,6694
38,00	11,00	2,00	12,00	31,7559

7.8   Conclusion

7.8.1   Summaries of Traditional Regressions

With traditional linear regression the old treatment was a borderline significant predictor at $p = 0,048$ of the novel treatment. More statistical power was desirable.

Neither was the overall R Square value of 0,219 a strong predictor (21,9% certainty about the outcome knowing the predictors. R square values under 25% are considered to be weak predictors. A problem with robust regressions in GENLIN is that no R Square values are provided, and, so, the strength cannot be precisely tested against kernel ridge models. Quantile regression does not provide R Square values, but pseudo R Square values can do the job instead (see the Chap. 5 entitled "Some Terminology"). And, so, its datafit can somewhat better be compared against kernel ridge regression and traditional linear regression. The best fit pseudo R Square values were with the quantiles 0,1, 0,2, 0,3,and the respective values 0,310, 0.259, and 0,220. All of them were slightly "betterfit" than the traditional regression R Square of 0,219. Yet all them were pretty poor predictors giving only 31,1%, 25,9%, and 22,0% certainty about the outcome knowing the predictors, and stronger prediction models are welcome. Kernel ridge regressions will be applied for the purpose next.

7.8.2 Summaries of Kernel Ridge Regressions

Obviously, the polynomial kernel density provided the best fit R Square value of no less than 0,831 (83,1% certainty is predicted by the predictors about the outcome. The scatterplot of the predicted newtreat values versus predicted ones shows a very nice linear pattern of the polynomial kernel ridge regression, while the spread of the residuals had a constant pattern.

7.9 References

All of the chapters of the current edition start with a brief review of the traditional analytic method of the different regression methods prior to the review of the relevant kernel ridge regression method. For the purpose, generally, data examples are used from the recent edition "Regression Analyses in Clinical Research for Starters and 2nd Levelers 2nd Edition, Springer Heidelberg Germany 2021", by the same authors. For a better understanding of differences between traditional and kernel ridge regressions, readers may benefit from the study of this edition first.

To readers requesting still more background, theoretical and mathematical information of computations given, several textbooks complementary to the current production and written by the same authors are available: Statistics applied to clinical studies 5th edition, 2012, Machine learning in medicine a complete overview 2nd edition, 2020, SPSS for starters and 2nd levelers 2nd edition, 2015, Clinical data analysis on a pocket calculator 2nd edition, 2016, Understanding clinical data analysis from published research, 2016, all of them edited by Springer Heidelberg Germany.

Chapter 8
Effect of Gene Expressions on Drug Efficacy, 250 Patients, Traditional Regressions vs Kernel Ridge Regression

Abstract In a 250 patient data file 12 highly expressed genes were tested for their effect on a clustered outcome variable of drug efficacy. The multiple variables regression showed that 6 genes were very significant independent predictors of drug efficacy scores. Kernel ridge regressions provided quite better datafit particularly

the Additive Chi2 kernel density model (R Square 0,766),
the Laplacian kernel density model (R Square 0,793),
the Polynomial kernel density model (R Square 0,951).

Much better statistics can thus be obtained with the help of kernel ridge regressions using one of the eight kernel density models available in the kernel ridge regression menu of SPSS statistical software.

Keywords Kernel ridge regression · Additive Chi2 kernel density model. · Laplacian kernel density model · Polynomial kernel density model

8.1 Summary

In a 250 patient data file 12 highly expressed genes were tested for effect on a clustered outcome variable of drug efficacy. The multiple variables regression showed that 6 genes were very significant independent predictors of drug efficacy scores. When tested against kernel ridge linear regression, the results of the traditional multiple variables regression and the latter regressions were pretty much similar. Using other kernel density models provided quite better datafit like the Additive Chi2 kernel density, the Laplacian kernel density, and the Polynomial kernel density models. Better statistics can thus be obtained with the help of kernel

---

Supplementary Information The online version contains supplementary material available at [https://doi.org/10.1007/978-3-031-10717-7_8].

ridge regressions using one of the eight kernel density models available in the kernel ridge regression menu of SPSS statistical software.

8.1.1 Summaries of Traditional Regressions

The R Square value of the traditional linear regression is 0,729. It means that we have 72,9% certainty about the summaryoutcome knowing the predictors, and we are only 27,1% uncertain. This is a very good predictive result. R Squares under 0,25% are very poor, 0,25–50% are reasonable, and over 0,50% are strong predictive models. Yet we may want to assess the predictive potential of the kernel ridge regression models.

8.1.2 Summaries of Kernel Ridge Regressions

The underneath table and graph are in the SPSS output. The R Square value of the Linear density model is 0,723, pretty good but not better than the R Square value from the traditional linear regression which was 0, 729 (adjusted 0,715).

Better datafits might be obtained by alternative kernel density models. To answer this question, analyses with more kernel density models were performed.

Additive Chi2 kernel density model	R Square = 0,766
Chi2 kernel density model	R Square = −0,028
Cosine kernel density model	R Square = 0,445
Laplacian kernel density model	R Square = 0,793
Polynomial kernel density model	R Square = 0,951
Radial basis neural network density function	R Square = −0,061
Sigmoid kernel density model	R Square = 0,000

Three kernel density models provided better fit data statistics with R Square values from 0,766 to 0,951. And so, if you need a data result better than that of the linear density kernel ridge model, you may wish to use one of those in your analysis.

8.2 Introduction

A 250 patient data file includes 27 variables consistent of both patients' microarray gene expression levels and their drug efficacy scores. All of the variables were standardized by scoring them on 11 points linear scales (0–10). The 250 patients' data-file was supposed to include 27 variables consistent of both patients' microarray gene expression levels and their drug efficacy scores. The following genes were highly expressed: the genes (Gs) 1–4, 16–19, and 24–27. Four variables were supposed to represent drug efficacy scores and were clustered as the outcome variables (O 1–4).

8.3   Data Example

The data from the first 13 patients are shown underneath. The entire data file entitled "genes" is in SpringerLink supplementary files.

G = gene.
O = outcome.

G 1	G 2	G 3	G 4	G 16	G 17	G 18	G 19	G 24	G 25	G 26	G 27	O 1	O 2	O 3	O 4
8	8	9	5	7	10	5	6	9	9	6	6	6	7	6	7
9	9	10	9	8	8	7	8	8	9	8	8	8	7	8	7
9	8	8	8	8	9	7	8	9	8	9	9	9	8	8	8
8	9	8	9	6	7	6	4	6	6	5	5	7	7	7	6
10	10	8	10	9	10	10	8	8	9	9	9	8	8	8	7
7	8	8	8	8	7	6	5	7	8	8	7	7	6	6	7
5	5	5	5	5	6	4	5	5	6	6	5	6	5	6	4
9	9	9	9	8	8	8	8	9	8	3	8	8	8	8	8
9	8	9	8	9	8	7	7	7	7	5	8	8	7	6	6
10	10	10	10	10	10	10	10	10	8	8	10	10	10	9	10
2	2	8	5	7	8	8	8	9	3	9	8	7	7	7	6
7	8	8	7	8	6	6	7	8	8	8	7	8	7	8	8
8	9	9	8	10	8	8	7	8	8	9	9	7	7	8	8

8.4   Traditional Linear Regression

The data from the first 13 patients are shown above. The entire data file entitled "genes" is in SpringerLink supplementary files. It is previously used by the authors in Machine learning in medicine a complete overview, Chap. 22, Springer Heidelberg Germany, 2016. A multiple variables linear regression was performed with the summary of the four drug efficacy scores as outcome variable and the 12 gene expression levels as covariates. Start by opening the data file in your computer with SPSS statistical software installed. We can now use the Menu commands.

Command:

Analyze....Regression....Dependent   Variables: enter   summaryoutcome.... Covariates: enter the 12 genes....click OK.

The underneath tables are in the output sheets. It shows that the genes 16, 17, 19, 24, 26, and 27 were significant predictors of the summaryoutcome drug efficacy.

Model Summary

Model	R	R Square	Adjusted R Square	Std. Error of the Estimate
1	,854[a]	,729	,715	3,64176

a. Predictors: (Constant), genetwentyseven, geneone, genesixteen, genetwentyfive, geneeighteen, genefour, genenineteen, genetwentyfour, genethree, geneseventeen, genetwentysix, genetwo

ANOVA[b]

Model		Sum of Squares	df	Mean Square	F	Sig.
1	Regression	8446,257	12	703,855	53,071	,000[a]
	Residual	3143,187	237	13,262		
	Total	11589,444	249			

a. Predictors: (Constant), genetwentyseven, geneone, genesixteen, genetwentyfive, geneeighteen, genefour, genenineteen, genetwentyfour, genethree, geneseventeen, genetwentysix, genetwo
b. Dependent Variable: summaryoutcome

Coefficients[a]

Model		Unstandardized Coefficients B	Std. Error	Standardized Coefficients Beta	t	Sig.
1	(Constant)	3,293	1,475		2,232	,027
	geneone	-,122	,189	-,030	-,646	,519
	genetwo	,287	,225	,078	1,276	,203
	genethree	,370	,228	,097	1,625	,105
	genefour	,063	,196	,014	,321	,748
	genesixteen	,764	,172	,241	4,450	,000
	geneseventeen	,835	,198	,221	4,220	,000
	geneeighteen	,088	,151	,027	,580	,563
	genenineteen	,576	,154	,188	3,751	,000
	genetwentyfour	,403	,146	,154	2,760	,006
	genetwentyfive	,028	,141	,008	,198	,843
	genetwentysix	,320	,142	,125	2,250	,025
	genetwentyseven	-,275	,133	-,092	-2,067	,040

a. Dependent Variable: summaryoutcome

The R Square value of the traditional linear regression is 0,729. It means that we have 72,9% certainty about the summaryoutcome knowing the predictors, and we are only 27,1% uncertain. This is a very good predictive result. R Squares under 0,25 are very poor, 0,25–0,50 are reasonable, and over 0,50 are strong predictive models. Yet we may want to assess the predictive potential of the kernel ridge regression models.

8.5 Kernel Ridge Regression

For the use of the menu kernel ridge regression your computer must be mounted with SPSS statistical software 2022, version 18.0.1.0. Kernel ridge regression methodology is also available through statistical software SAS, R, Matlab and more.

In SPSS 18.0.1.0 Command:

Menu....Analyze....Regression....Kernel Ridge Regression....Dependent: summaryoutcome....Covariates: enter the 12 genes....click Linear....click Options....mark Observed vs. Predicted....Mark Predicted ....click Continue.... click OK.

The underneath table and graph are in the SPSS output. The R Square value of the Linear density model is 0,723, pretty good but not better than the R Square value from the traditional linear regression which was 0, 729 (adjusted 0,715).

Model Summary[a,b]

Kernel	Alpha	R Square
Linear	1,000	,723

a. Dependent Variable: summaryoutcome

b. Model: geneone, genetwo, genethree, genefour, genesixteen, geneseventeen, geneeighteen, genenineteen, genetwentyfour, genetwentyfive, genetwentysix, genetwentyseven

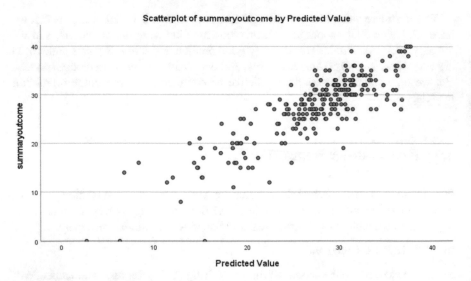

The above graph shows a good linear pattern.

Returning to the SPSS dataview screen, it can be observed, that SPSS has added a column of predicted values, the latter column. They are the predicted values of the observed summaryoutcomes. Only the first 10 patients of the 250 are in the underneath table. The are pretty close to the original values, and, so, the kernel ridge regression, obviously, seems to be a pretty close predictor of the originally measured summaryoutcomes.

summaryoutcome	predicted value
26,00	29,8284
30,00	1,2746
33,00	1,1674
27,00	24,4795
31,00	33,7631
26,00	27,1160
21,00	19,9343
32,00	29,7246
27,00	29,1634
39,00	36,6644

We may ask, whether better datafits might be obtained by alternative kernel density models. In order to answer this question, we will perform analyses with more kernel density models. For the purpose similar commands must be given. The underneath table shows the results.

Additive Chi2 kernel density model	R Square = 0,766
Chi2 kernel density model	R Square = −0,028
Cosine kernel density model	R Square = 0,445
Laplacian kernel density model	R Square = 0,793
Polynomial kernel density model	R Square = 0,951
Radial basis neural network density function	R Square = −0,061
Sigmoid kernel density model	R Square = 0,000

Three kernel density models provided better fit data statistics with R Square values from 0,766 to 0,951. And so, if you need a data result better than that of the linear density kernel ridge model, you may wish to use one of those in your analysis. Note that negative R Squares are sometimes produced. They express very poor data fit. Negative R Squares are explained in the Chap. 5 entitled "Some Terminology".

8.6 Conclusion

8.6.1 Summaries of Traditional Regressions

The R Square value of the traditional linear regression is 0,729. It means that we have 72,9% certainty about the summaryoutcome knowing the predictors, and we are only 27,1% uncertain. This is a very good predictive result. R Squares under 0,25 are very poor, 0,25–50 are reasonable, and over 0,50 are strong predictive models. Yet we may want to assess the predictive potential of the kernel ridge regression models.

8.6.2 Summaries of Kernel Ridge Regressions

The underneath table and graph are in the SPSS output. The R Square value of the Linear density model is 0,723, pretty good but not better than the R Square value from the traditional linear regression which was 0, 729 (adjusted 0,715).

Better datafits might be obtained by alternative kernel density models. To answer this question, analyses with more kernel density models were performed.

Additive Chi2 kernel density model	R Square = 0,766
Chi2 kernel density model	R Square = −0,028
Cosine kernel density model	R Square = 0,445
Laplacian kernel density model	R Square = 0,793
Polynomial kernel density model	R Square = 0,951

Radial basis neural network density function   R Square = −0,061
Sigmoid kernel density model   R Square = 0,000

Three kernel density models provided better fit data statistics with R Square values from 0,766 to 0,951. And so, if you need a data result better than that of the linear density kernel ridge model, you may wish to use one of those in your analysis.

8.7   References

All of the chapters of the current edition start with a brief review of the traditional analytic method of the different regression methods prior to the review of the relevant kernel ridge regression method. For the purpose, generally, data examples are used from the recent edition "Regression Analyses in Clinical Research for Starters and 2nd Levelers 2nd Edition, Springer Heidelberg Germany 2021", by the same authors. For a better understanding of differences between traditional and kernel ridge regressions, readers may benefit from the study of this edition first.

To readers requesting still more background, theoretical and mathematical information of computations given, several textbooks complementary to the current production and written by the same authors are available: Statistics applied to clinical studies 5th edition, 2012, Machine learning in medicine a complete overview 2nd edition, 2020, SPSS for starters and 2nd levelers 2nd edition, 2015, Clinical data analysis on a pocket calculator 2nd edition, 2016, Understanding clinical data analysis from published research, 2016, all of them edited by Springer Heidelberg Germany.

Chapter 9
Effect of Gender, Treatment, and Their Interactions on Numbers of Paroxysmal Atrial Fibrillations, 40 Patients, Traditional Regressions vs Kernel Ridge Regression

Abstract In a 40 patient study the effect of gender, treatment modalities, and their interaction was studied on the prevention of episodes of paroxysmal atrial fibrillation (paf). The R Square of the traditional regression with two predictors was 0,336 (33,6% certainty of prediction). Kernel ridge regressions were much more powerful.

Chi2 kernel density model	R Square 0,654
Laplacian	R Square 0,641
Polynomial	R Square 0,716
RBF (radial basis function)	R Square 0,641.

The R Square of the traditional regression with three predictors was 0,758 (75,8% of certainty about the outcome is given. This is much better than the 33,6% prediction of the two predictor traditional linear regression. Kernel ridge regressions with three predictors remained very powerful.

Additive_chi2 kernel density model	R Square 0,710
Chi2	R Square 0,634
Laplacian	R Square 0,680
Radial basis function	R Square 0,680
Polynomial	R Square 0,722.

Keywords Kernel ridge regression · Additive_chi2 kernel densitymodel · Chi2 · Kernel density model · Laplacian kernel density model · Radial basis function · Kernel density model · Polynomial kernel density model

Supplementary Information The online version contains supplementary material available at [https://doi.org/10.1007/978-3-031-10717-7_9].

9.1 Summary

In a 40 patient study the effect of (1) gender, (2) treatment modalities, and (3) the interaction of the two on the prevention of episodes of paroxysmal atrial fibrillation (paf) was studied. Interaction adjusted multiple variables regression predicted significant effects of gender and interaction on the outcome "prevention of paroxysmal atrial fibrillation at $p = 0,0001$. The predictor treatment modality (metroprolol or verapamil) was not statistically significant.

9.1.1 Summaries of Two Predictor Regressions

The R Square was 0,336 meaning that the outcome is predicted by the two predictors gender and treatment with 33,6% certainty, which is a reasonable but not strong predictive power. More power is welcome. Therefore, kernel ridge regressions were performed.

Four kernel density models produced R Square values (much) larger than that of the traditional linear regression (0,336).

Chi2   R Square 0,654
Laplacian   R Square 0,641
Polynomial   R Square 0,716
RBF (radial basis function)   R Square 0,641.

With the polynomial kernel density model 71,6% certainty about the outcome is given.

Obviously the kernel ridge model provided a much better R Square value and thus better fit for the data.

9.1.2 Summaries of Three Predictor Regressions

By three predictors 75,8% of certainty about the outcome is given by the traditional linear regression. This is much better than the 33,6% prediction of the two predictor traditional linear regression. The polynomial kernel ridge density model provided the best data fit out of eight alternative kernel density models. An R Square of 72.2% is very good and virtually equally powerful as that of the traditional linear model with 75,8%, although not better.

Five kernel density models produced R Square values (much) larger than that of the traditional two predictor regression (0,336).

Additive_chi2	R Square 0,710
Chi2	R Square 0,634
Laplacian	R Square 0,680
Radial basis function	R Square 0,680
Polynomial	R Square 0,722.

9.2 Introduction

In a 40 patient study the effect of (1) gender, (2) treatment modalities, and (3) the interaction of the two on the prevention of episodes of paroxysmal atrial fibrillation (paf) was studied. Interaction adjusted multiple variables regression predicted significant effects of gender and interaction on the outcome "prevention of paroxysmal atrial fibrillation at $p = 0,0001$. The predictor treatment modality (metroprolol or verapamil) was not statistically significant.

9.3 Data Example

In a 40 patient study the effect of (1) gender, (2) treatment modalities, and (3) the interaction of the two on the prevention of episodes of paroxysmal atrial fibrillation was studied. The first 10 patients of 40 patient study is underneath. The data file is in SpringerLink supplementary files, and is entitled "gender x treatment".

paf	gender	treatment	interaction
52,00	1,00	,00	,00
48,00	1,00	,00	,00
43,00	1,00	,00	,00
50,00	1,00	,00	,00
43,00	1,00	,00	,00
44,00	1,00	,00	,00
46,00	1,00	,00	,00
46,00	1,00	,00	,00
43,00	1,00	,00	,00
49,00	1,00	,00	,00

paf = numbers of paroxysmal fibrillations per patients
gender = 1 = male, 0 = female
treatment = either verapamil or metroprolol (verapamil = 1, metroprolol = 0)
interaction = gender x treatment

9.4   Traditional Linear Regression with Two Predictors

In your computer mounted with SPSS statistical software dowwnload the above datafile.

Command:

Analyze....Regression....Linear....Dependent: enter paf....Independent(s): enter gender, treatment (metro-vera)....click OK.

The underneath tables are in the output sheets.

Model Summary

Model	R	R Square	Adjusted R Square	Std. Error of the Estimate
1	,579[a]	,336	,300	5,60598

a. Predictors: (Constant), metro-vera, gender

ANOVA[b]

Model		Sum of Squares	df	Mean Square	F	Sig.
1	Regression	587,600	2	293,800	9,349	,001[a]
	Residual	1162,800	37	31,427		
	Total	1750,400	39			

a. Predictors: (Constant), metro-vera, gender
b. Dependent Variable: paf

Coefficients[a]

Model		Unstandardized Coefficients		Standardized Coefficients	t	Sig.
		B	Std. Error	Beta		
1	(Constant)	41,100	1,535		26,771	,000
	gender	1,000	1,773	,076	,564	,576
	metro-vera	-7,600	1,773	-,574	-4,287	,000

a. Dependent Variable: paf

The R Square is 0,336 which means that the outcome is predicted by the two predictors gender and treatment with 33,6% certainty, a reasonable but not strong predictive power. More power would be welcome. Therefore, kernel ridge regressions will be performed.

9.5   Kernel Ridge Regression with Two Predictors

In your computer mounted with SPSS statistical software 2022 version 28.0.1.0 downlaod the datafile, and command.

Command:

Analyze....Regression....Kernel Ridge Regression....Dependent: enter paf....Independent(s): enter gender, treatment....click Linear....click OK.
    The output is below.

Model Summary[a,b]

Kernel	Alpha	R Square
Linear	1,000	-12,554

a. Dependent Variable: paf

b. Model: gender, treat

Obviously, thc linear kernel density model has an extremely poor fit for the data and more kernel density models are requested. The phenomenon of negative R Square values is pretty common and is explained in the Chap. 5 entitled "Some Terminology".

Command:

Analyze....Regression....Kernel Ridge Regression....Dependent: enter paf....Independent(s): enter gender, treatment....click Linear....click Additive_chi2....click OK.

Model Summary[a,b]

Kernel	Alpha	R Square
Additive_chi2	1,000	,313

a. Dependent Variable: paf

b. Model: gender, treat

The Additive_Chi2 kernel fits much better, and subsequent kernel density models are similarly requested.

Model Summary[a,b]

Kernel	Alpha	Gamma	R Square
Chi2	1,000	1,000	,654

a. Dependent Variable: paf

b. Model: gender, treat

Model Summary[a,b]

Kernel	Alpha	R Square
Cosine	1,000	-9,491

a. Dependent Variable: paf
b. Model: gender, treat

Model Summary[a,b]

Kernel	Alpha	Gamma	R Square
Laplacian	1,000	,500	,641

a. Dependent Variable: paf
b. Model: gender, treat

Model Summary[a,b]

Kernel	Alpha	Gamma	Coef0	Degree	R Square
Polynomial	1,000	,500	1,000	3,000	,716

a. Dependent Variable: paf
b. Model: gender, treat

Model Summary[a,b]

Kernel	Alpha	Gamma	R Square
RBF	1,000	,500	,641

a. Dependent Variable: paf
b. Model: gender, treat

Model Summary[a,b]

Kernel	Alpha	Gamma	Coef0	R Square
Sigmoid	1,000	,500	1,000	-,454

a. Dependent Variable: paf
b. Model: gender, treat

Four kernel density models produced R Square values (much) larger than that of the traditional linear regression (0,336).

Chi2	R Square 0,654
Laplacian	R Square 0,641

Polynomial   R Square 0,716
RBF (radial basis function)   R Square 0,641.

With the polynomial kernel density model no less than 71,6% certainty about the outcome is given by the two predictor model. Obviously the kernel ridge model provided a much better R Square value and thus better fit for the data. No less than 71,6% of certainty about the outcome is given by the predictors.

9.6   Traditional Linear Regression with Three Predictors

In your computer mounted with SPSS statistical software command:

Analyze....Regression....Linear....Dependent: enter paf....Independent(s): enter gender, treatment, interaction....click OK.

The underneath table are in the output.

Model Summary

Model	R	R Square	Adjusted R Square	Std. Error of the Estimate
1	,871[a]	,758	,738	3,42864

a. Predictors: (Constant), interaction, metro-vera, gender

ANOVA[a]

Model		Sum of Squares	df	Mean Square	F	Sig.
1	Regression	1327,200	3	442,400	37,633	,000[b]
	Residual	423,200	36	11,756		
	Total	1750,400	39			

a. Dependent Variable: paf
b. Predictors: (Constant), interaction, metro-vera, gender

Coefficients[a]

Model		Unstandardized Coefficients		Standardized Coefficients	t	Sig.
		B	Std. Error	Beta		
1	(Constant)	36,800	1,084		33,941	,000
	gender	9,600	1,533	,726	6,261	,000
	metro-vera	1,000	1,533	,076	,652	,518
	interaction	-17,200	2,168	-1,126	-7,932	,000

a. Dependent Variable: paf

Obviously 75,8% of certainty about the outcome is given simultaneously by the three predictors. This is much better than the 33,6% prediction of the two predictor traditional linear regression.

9.7   Kernel Ridge Regression with Three Predictors

Next, a three predictor kernel ridge regression analysis will be performed. For the purpose download the datafile in a computer installed with SPSS statistical software 2022 version 0.1.0. Open the file and start commanding.

Command:

Analyze....Regression....Kernel ridge regression....Dependent: enter paf....Independent(s): enter gender, treatment, interaction....Linear....click OK.

Model Summary[a,b]

Kernel	Alpha	R Square
Linear	1,000	-7,633

a. Dependent Variable: paf

b. Model: gender, treat, interaction

Again the linear kernel ridge regression has a very poor datafit. More kernel density models are applied giving similar commands.

Model Summary[a,b]

Kernel	Alpha	R Square
Additive_chi2	1,000	,710

a. Dependent Variable: paf

b. Model: gender, treat, interaction

Model Summary[a,b]

Kernel	Alpha	Gamma	R Square
Chi2	1,000	1,000	,634

a. Dependent Variable: paf

b. Model: gender, treat, interaction

Model Summary[a,b]

Kernel	Alpha	R Square
Cosine	1,000	-7,349

a. Dependent Variable: paf

b. Model: gender, treat,
   interaction

Model Summary[a,b]

Kernel	Alpha	Gamma	R Square
Laplacian	1,000	,333	,680

a. Dependent Variable: paf

b. Model: gender, treat, interaction

Model Summary[a,b]

Kernel	Alpha	Gamma	Coef0	Degree	R Square
Polynomial	1,000	,333	1,000	3,000	,722

a. Dependent Variable: paf

b. Model: gender, treat, interaction

Model Summary[a,b]

Kernel	Alpha	Gamma	R Square
RBF	1,000	,333	,680

a. Dependent Variable: paf

b. Model: gender, treat, interaction

Model Summary[a,b]

Kernel	Alpha	Gamma	Coef0	R Square
Sigmoid	1,000	,333	1,000	,099

a. Dependent Variable: paf

b. Model: gender, treat, interaction

The polynomial kernel ridge density model provided the best data fit out of eight alternative kernel density models. An R Square of 72,2% is very good and virtually equally powerful as that of the traditional linear model with 75,8%, although not better.

Five kernel density models produced R Square values (much) larger than that of the traditional linear regression (0,336).

Additive_chi2   R Square 0,710
Chi2   R Square 0,634
Laplacian   R Square 0,680
Radial basis function   R Square 0,680
Polynomial   R Square 0,722.

9.8   Conclusion

In a 40 patient study the effect of (1) gender, (2) treatment modalities, and (3) the interaction of the two on the prevention of episodes of paroxysmal atrial fibrillation (PAF) was studied.

9.8.1   Summaries of Two Predictor Regressions

The R Square was 0,336 meaning that the outcome is predicted by the two predictors gender and treatment with 33,6% certainty, which is a reasonable but not strong predictive power. More power is welcome. Therefore, kernel ridge regressions were performed.

Four kernel density models produced R Square values (much) larger than that of the traditional linear regression (0,336).

Chi2   R Square 0,654
Laplacian   R Square 0,641
Polynomial   R Square 0,716
RBF (radial basis function)   R Square 0,641.

With the polynomial kernel density model 71,6% certainty about the outcome is given.

Obviously the kernel ridge model provided a much better R Square value and thus better fit for the data.

9.8.2   Summaries of Three Predictor Regressions

By three predictors 75,8% of certainty about the outcome is given by the traditional linear regression. This is much better than the 33,6% prediction of the two predictor traditional linear regression. The polynomial kernel ridge density model provided the best data fit out of eight alternative kernel density models. An R Square of 72.2%

is very good and virtually equally powerful as that of the traditional linear model with 75,8%, although not better.

Five kernel density models produced R Square values (much) larger than that of the traditional two predictor regression (0,336).

Additive_chi2   R Square 0,710
Chi2   R Square 0,634
Laplacian   R Square 0,680
Radial basis function R Square 0,680
Polynomial   R Square 0,722.

9.9 References

All of the chapters of the current edition start with a brief review of the traditional analytic method of the different regression methods prior to the review of the relevant kernel ridge regression method. For the purpose, generally, data examples are used from the recent edition "Regression Analyses in Clinical Research for Starters and 2nd Levelers 2nd Edition, Springer Heidelberg Germany 2021", by the same authors. For a better understanding of differences between traditional and kernel (ridge) regressions, readers may benefit from the study of this edition first.

To readers requesting still more background, theoretical and mathematical information of computations given, several textbooks complementary to the current production and written by the same authors are available: Statistics applied to clinical studies 5th edition, 2012, Machine learning in medicine a complete overview 2nd edition, 2020, SPSS for starters and 2nd levelers 2nd edition, 2015, Clinical data analysis on a pocket calculator 2nd edition, 2016, Understanding clinical data analysis from published research, 2016, all of them edited by Springer Heidelberg Germany.

Chapter 10
Effect of Laboratory Predictors on Septic Mortality, 200 Patients, Traditional Regression vs Kernel Ridge Regression

Abstract In a 200 patient data file of patients with sepsis the effect of laboratory predictors on survival/septic death was assessed. Traditional regression consisted of binary logistic regression with the logodds of survival from sepsis as outcome and the various laboratory values as predictors. Obviously, bilirubine, c-reactive protein, leucos are significant predictors, while the remainder of the predictors were insignificant. This result was pretty disappointing, because clinically we would expect many more laboratory values to predict survival from sepsis. In a kernel ridge regression (Cosine density model) the R Square was better sensitive than the Cox and Snell pseudo R Square: 0, 811 vs 0,700. The cosine frequency distribution can be observed in the Chap. 5 entitled "Some Terminology".

Keywords Binary logistic regression · Kernel ridge regression · Cosine kernel density model

10.1 Summary

Clinical chemistry has been recognized as helpful as a predictor of morbidity/mortality scores. In a 200 patient data file of patients with sepsis the effect of laboratory predictors on survival/septic death was assessed. Traditional regression consisted of binary logistic regression with the logodds of survival from sepsis as outcome and the various laboratory values as predictors. Obviously, bilirubine, c-reactive protein, leucos are significant predictors, while the remainder of the predictors were insignificant. This result was pretty disappointing, because clinically we would expect many more laboratory values to predict survival from sepsis. In a kernel ridge regression (Cosine density model) the R Square was better sensitive than the Cox and Snell pseudo R Square: 0, 811 vs 0,700. The cosine frequency distribution can be observed in the Chap. 5 entitled "Some Terminology".

Supplementary Information The online version contains supplementary material available at [https://doi.org/10.1007/978-3-031-10717-7_10].

10.1.1   Summaries of the Traditional Regressions

The test–retest reliability of the manifest variables as assessed with Cronbach's alphas using the correlation coefficients with one variable missing are given. All of the missing data files should produce at least by 80% the same result as those of the non-missing data files (alphas >80%). Indeed, none of the predictor variables after deletion reduced the test–retest reliability. The data are reliable. A binary logistic regression can be performed.

In order to compare the binary outcome logistic model with the kernel ridge model, it would be convenient to compare the R Square values of either of the two methods. However, the logistic models do not provide the maths to produce R Square values. Instead, however, Cox and Snell (1989) proposed as alternative a pseudo R Square based on likelihood statistics. Its upperbound is like that of the true R Square 1,000. However, It sometimes underestimates the certainty proportion in the binary data as given. In our data example the Cox and Snell pseudo R Square equalled 0,700.

10.1.2   Summaries of the Kernel Ridge Regressions

The Cosine kernel density model produced an R Square somewhat betterfit than that of the Linear kernel density model, and it was actually the bestfit model as compared with the traditional binary logistic model, 0,811 versus 0, 700.

10.2   Introduction

Clinical chemistry has been recognized as helpful as a predictor of morbidity/mortality scores. In a 200 patient data file of patients with sepsis the effect of laboratory predictors on survival/septic death was assessed. Traditional regression analysis consisted of binary logistic regression with the logodds of survival from sepsis as outcome and the various laboratory values as predictors.

10.3   Data Example

In a 200 patient data file of patients with sepsis the effect of laboratory predictors on survival/septic death was assessed. Traditional consisted of binary logistic regression with the logodds of survival from sepsis as outcome and the various laboratory values as predictors.

Variables

1	2	3	4	5	6	7	8	9	10	11
sv	gam	asat	alat	bili	ureum	creat	creat-cl	esr	c-react	leuc
,00	20,00	23,00	34,00	2,00	3,40	89,00	-111,00	2,00	2,00	5,00
,00	14,00	21,00	33,00	3,00	2,00	67,00	-112,00	7,00	3,00	6,00
,00	30,00	35,00	32,00	4,00	5,60	58,00	-116,00	8,00	4,00	4,00
,00	35,00	34,00	40,00	4,00	6,00	76,00	-110,00	6,00	5,00	7,00
,00	23,00	33,00	22,00	4,00	6,10	95,00	-120,00	9,00	6,00	6,00
,00	26,00	31,00	24,00	3,00	5,40	78,00	-132,00	8,00	4,00	8,00
,00	15,00	29,00	26,00	2,00	5,30	47,00	-120,00	12,00	5,00	5,00
,00	13,00	26,00	24,00	1,00	6,30	65,00	-132,00	13,00	6,00	6,00
,00	26,00	27,00	27,00	4,00	6,00	97,00	-112,00	14,00	6,00	7,00
,00	34,00	25,00	13,00	3,00	4,00	67,00	-125,00	15,00	7,00	6,00

sv = survival / septic death
gam = gamma-gt
bili = bilirubine
creat = creatinine
creat-cl = creatinine clearance
esr = erythrocyte sedimentation rate
c-react = c-reactive protein
leuc = leucocyte count

As a data example, a data set of 200 patients at risk of septic death is used. The first 10 patients are in the table above. The individual patient data are given in the SPSS data file entitled "lab", and is in SpringerLink supplementary files. It is previously used by the authors in Machine learning in medicine, Chap. 14, Springer Heidelberg Germany, 2013.

10.4 Test and Retest Reliability

We will first test the test–retest reliability of the original variables. The test–retest reliability of the original variables should be assessed with Cronbach' s alphas using the correlation coefficients after deletion of one variable: all of the data files should produce by at least 80% the same result as that of the non-deleted data file (alphas >80%). Open the data file in your computer installed with SPSS statistical software.

Command:

Analyze. . . .Scale. . . .Reliability Analysis. . . .transfer original variables to Variables box. . . .click Statistics. . . .mark Scale if item deleted. . . .mark Correlations. . . . Continue . . . .OK.

The underneath table is in the SPSS output sheets. The test–retest reliability of the manifest variables as assessed with Cronbach's alphas using the correlation coefficients with one variable missing are given. All of the missing data files should produce at least by 80% the same result as those of the non-missing data files (alphas >80%).

Item-Total Statistics

	Scale Mean if Item Deleted	Scale Variance if Item Deleted	Corrected Item-Total Correlation	Squared Multiple Correlation	Cronbach's Alpha if Item Deleted
gammagt	650,6425	907820,874	,892	,827	,805
asat	656,7425	946298,638	,866	,772	,807
alat	660,5975	995863,808	,867	,826	,803
bili	789,0475	1406028,628	,938	,907	,828
ureum	835,8850	1582995,449	,861	,886	,855
creatinine	658,3275	1244658,421	,810	,833	,814
creatinine clearance	929,7875	1542615,450	,721	,688	,849
esr	812,2175	1549217,863	,747	,873	,850
c-reactive protein	827,8975	1590791,528	,365	,648	,857
leucos	839,0925	1610568,976	,709	,872	,859

The table shows, that, indeed, none of the predictor variables after deletion reduced the test–retest reliability. The data are reliable. A binary logistic regression was performed.

10.5   Binary Logistic Regression

Command:

Analyze....Regression....Binary Logistic....Dependent: survival sepsis....Covariates: enter all of the laboratory predictors....click OK.

Variables in the Equation

		B	S.E.	Wald	df	Sig.	Exp(B)
Step 1[a]	gammagt	,020	,019	1,141	1	,285	1,020
	asat	,003	,017	,034	1	,854	1,003
	alat	,001	,015	,002	1	,965	1,001
	bili	,038	,017	5,237	1	,022	1,039
	ureum	-,172	,112	2,350	1	,125	,842
	creatinine	,001	,008	,016	1	,901	1,001
	creatinine clearance	-,036	,034	1,147	1	,284	,964
	esr	-,021	,049	,187	1	,665	,979
	c-reactive protein	-,068	,033	4,270	1	,039	,934
	leucos	1,236	,395	9,781	1	,002	3,442
	Constant	-17,739	6,609	7,205	1	,007	,000

a. Variable(s) entered on step 1: gammagt, asat, alat, bili, ureum, creatinine, creatinine clearance, esr, c-reactive protein, leucos.

Model Summary

Step	-2 Log likelihood	Cox & Snell R Square
1	35,621[a]	,700

a. Estimation terminated at iteration number 10 because parameter estimates changed by less than ,001.

The above tables are in the output sheets. Obviously, bilirubine, c-reactive protein, leucos are significant predictors, while the remainder of the predictors were insignificant. This result is pretty disappointing, because, clinically, we would expect many more laboratory values to predict survival from sepsis.

In order to compare the binary outcome logistic model versus the kernel ridge model, it would be convenient to compare the R Square values of either of the two methods. However, the logistic models do not provide the maths to produce R Square values. Instead, however, Cox and Snell (1989) proposed as alternative a pseudo R Square based on likelihood statistics. Its upperbound is like that of the true R Square 1,000. However, It sometimes underestimates the certainty proportion in the binary data as given. In our data example the Cox and Snell pseudo R Square equals 0,700.

10.6   Kernel Ridge Regression

In a computer installed with SPSS statistical software 2022 version 28.0.1.0 a kernel ridge regression of the same data will be performed.

Command:

Menu....Analyze....Regression....Kernel Ridge Regression....Dependent: survival sepsis....Covariates: enter all of the laboratory predictors....click Linear....click Options....mark Observed vs. Predicted....Mark Predicted ....click Continue.... click OK.

The output sheets give an kernel ridge R Square value, and a plot of predicted values of these binary data. The R Square of the kernel ridge was no less than 0,768, which means that 76,8% of the certainty about the outcome survival from sepsis or not is predicted by the model.

Model Summary[a,b]

Kernel	Alpha	R Square
Linear	1,000	,768

a. Dependent Variable:
   VAR00001

b. Model: VAR00002,
   VAR00003, VAR00004,
   VAR00005, VAR00006,
   VAR00007, VAR00008,
   VAR00009, VAR00010,
   VAR00011

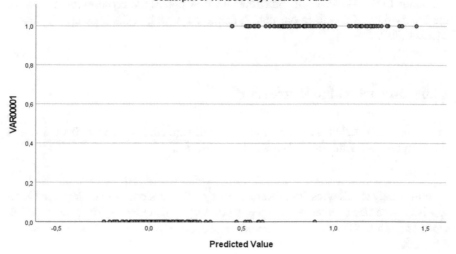

Scatterplot of VAR00001 by Predicted Value

The above scatterplot of predicted values versus the outcome variables is also produced by SPSS statistical software. The bimary logistic version of the kernel ridge analysis provides only two outcome values 0,0 and 1,0 and a number of different predicted values. The graph shows that a linear regression line can drawn between the mean of the plot of either of the two outcome values.

The R Square of the kernel ridge linear density model was 0,768. The R square was somewhat better sensitive than the pseudo R Square from the traditional binary logistic model = 0,700.

We will assess more kernel density models in order to find better data fit. For the purpose similar commands are given.

Model Summary[a,b]

Kernel	Alpha	R Square
Cosine	1,000	,811

a. Dependent Variable:
   VAR00001

b. Model: VAR00002,
   VAR00003, VAR00004,
   VAR00005, VAR00006,
   VAR00007, VAR00008,
   VAR00009, VAR00010,
   VAR00011

Model Summary[a,b]

Kernel	Alpha	Gamma	R Square
Laplacian	1,000	,100	,535

a. Dependent Variable: VAR00001

b. Model: VAR00002, VAR00003,
   VAR00004, VAR00005, VAR00006,
   VAR00007, VAR00008, VAR00009,
   VAR00010, VAR00011

Model Summary[a,b]

Kernel	Alpha	Gamma	R Square
RBF	1,000	,100	,533

a. Dependent Variable: VAR00001

b. Model: VAR00002, VAR00003,
   VAR00004, VAR00005, VAR00006,
   VAR00007, VAR00008, VAR00009,
   VAR00010, VAR00011

Model Summary[a,b]

Kernel	Alpha	Gamma	Coef0	R Square
Sigmoid	1,000	,100	1,000	,000

a. Dependent Variable: VAR00001

b. Model: VAR00002, VAR00003, VAR00004,
   VAR00005, VAR00006, VAR00007, VAR00008,
   VAR00009, VAR00010, VAR00011

The above tables show, that the Cosine kernel density model produced an R Square somewhat betterfit than that of the Linear kernel density model, and it was actually the bestfit model as compared with the traditional binary logistic model, 0,811 versus 0,768. The above kernel rdige regressions use five different kernel density models. The graphical presentations of all of them are in the Chap. 5 entitled "Some Terminology".

10.7   Conclusion

10.7.1   Summaries of the Traditional Regressions

The test–retest reliability of the manifest variables as assessed with Cronbach's alphas using the correlation coefficients with one variable missing are given. All of the missing data files should produce at least by 80% the same result as those of the non-missing data files (alphas >80%). Indeed, none of the predictor variables after deletion reduced the test–retest reliability. The data are reliable. A binary logistic regression can be performed.

In order to compare the binary outcome logistic model with the kernel ridge model, it would be convenient to compare the R Square values of either of the two methods. However, the logistic models do not provide the maths to produce R Square values. Instead, however, Cox and Snell (1989) proposed as alternative a pseudo R Square based on likelihood statistics. Its upperbound is like that of the true R Square 1,000. However, It sometimes underestimates the certainty proportion in the binary data as given. In our data example the Cox and Snell pseudo R Square equalled 0,700.

10.7.2   Summaries of the Kernel Ridge Regressions

The Cosine kernel density model produced an R Square somewhat betterfit than that of the Linear kernel density model, and it was actually the bestfit model as compared with the traditional binary logistic model, 0,811 versus 0, 700.

10.8   References

All of the chapters of the current edition start with a brief review of the traditional analytic method of the different regression methods prior to the review of the relevant kernel ridge regression method. For the purpose, generally, data examples are used from the recent edition "Regression Analyses in Clinical Research for Starters and 2nd Levelers 2nd Edition, Springer Heidelberg Germany 2021", by the same authors. For a better understanding of differences between traditional and kernel ridge regression regressions, readers may benefit from the study of this edition first.

To readers requesting still more background, theoretical and mathematical information of computations given, several textbooks complementary to the current production and written by the same authors are available: Statistics applied to clinical studies 5th edition, 2012, Machine learning in medicine a complete overview, 2015, SPSS for starters and 2nd levelers 2nd edition, 2015, Clinical data analysis on a pocket calculator 2nd edition, 2016, Understanding clinical data analysis from published research, 2016, all of them edited by Springer Heidelberg Germany.

Chapter 11
Effect of Month on Mean C-Reactive Protein, 18 Months, Traditional Regressions vs Kernel Ridge Regression

Abstract The effect of month on mean c-reactive protein levels was assessed in 18 subsequent months. Lag numbers are used for analysis, with monthly measures of paired differences between the first and subsequent monthly measures.

The significant positive autocorrelations at the month no. 13 (correlation coefficients of 0,42 (SE 0,14, t-value 3,0, p < 0,01) supports seasonality, and so does the pattern of partial autocorrelation coefficients (not shown): it gradually falls, and a partial autocorrelation coefficient of zero is observed 1 month after month 13. The strength of the seasonality is assessed using the magnitude of $r^2 = 0,42^2 = 0,18$. This would mean that the lag curve predicts the datacurve by only 18%, and, thus, that 82% is unexplained. And so, evidence of seasonality is pretty weak. Instead various curvilinear traditional regressions were assessed. A poor fit was shown for all of them. In fact, the best result was obtained by the S model with a p-value, however, of only 0,12 and an R square value of 0,106. Kernel ridge regression as obtained by he Additive Chi square density model provided a somewhat better result with an R square value of 0,131 (13,1% certainty).

Keywords Autocorrelations · Lag numbers · Curvilinear regressions · Additive Chi2 kernel density model

11.1 Summary

The effect of month on mean c-reactive protein levels was assessed in 18 subsequent months. Usually lag numbers are used for analysis, with, e.g., monthly measures meaning the paired differences between the first monthly measures as compared to subsequent monthly measures.

Supplementary Information The online version contains supplementary material available at [https://doi.org/10.1007/978-3-031-10717-7_11].

11.1.1   Summaries of Traditional Regressions

11.1.1.1   Autocorrelations

The monthly autocorrelation coefficients with their 95% confidence intervals show that the magnitude of the monthly autocorrelations changes sinusoidally. The significant positive autocorrelations at the month no.13 (correlation coefficients of 0,42 (SE 0,14, t-value 3,0, p < 0,01)) further supports seasonality, and so does the pattern of partial autocorrelation coefficients (not shown): it gradually falls, and a partial autocorrelation coefficient of zero is observed one month after month 13. The strength of the seasonality is assessed using the magnitude of $r^2 = 0,42^2 = 0,18$. This would mean that the lag curve predicts the datacurve by only 18%, and, thus, that 82% is unexplained. And so, the seasonality may be statistically significant, but it is pretty weak, and a lot of unexplained variability, otherwise called noise, is in these data.

11.1.1.2   Curvilinear Regressions

The effect of times on C-reactive protein was studied using traditional curvilinear regression. The traditional linear OLS regression provided a very poor fit for the C-reactive protein outcome data. Instead various curvilinear traditional regressions were assessed. A poor fit was shown for all of them. In fact, the best result was obtained by the S model with a p-value, however, of only 0,12 and an R square value of 0,106.

11.1.2   Summaries of Kernel Ridge Regressions

The best fit kernel density model was obtained by he Additive Chi square density model with an R square value of 0,131. This R square value was slightly better than that of the above traditional curvilinear regression S model with an p-value of 0,12 and an R square value of 0,106. This would mean that by the S model 10,6% of the certainty about the outcome was provided, while by the Additive chi square kernel density model 13,1% of the certainty about the outcome was provided.

11.2   Introduction

For analysis the statistical module Autocorrelations in the SPSS' module Forecasting is required. The data file is in SpringerLink supplementary files, and is entitled "seasonality". Start by opening the file in your computer mounted with SPSS statistical software 2022 Version 28 0.1.0. In this chapter traditional autocorrelation analysis and was tested against traditional curvilinear regressions and kernel ridge regressions of the same data.

11.2.1 Summaries of Traditional Regressions

The monthly autocorrelation coefficients with their 95% confidence intervals show that the magnitude of the monthly autocorrelations changes sinusoidally. The significant positive autocorrelations at the month no.13 (correlation coefficients of 0,42 (SE 0,14, t-value 3,0, p < 0,01)) further supports seasonality, and so does the pattern of partial autocorrelation coefficients (not shown): it gradually falls, and a partial autocorrelation coefficient of zero is observed one month after month 13. The strength of the seasonality is assessed using the magnitude of $r^2 = 0,42^2 = 0,18$. This would mean that the lag curve predicts the datacurve by only 18%, and, thus, that 82% is unexplained. And so, the seasonality may be statistically significant, but it is pretty weak, and a lot of unexplained variability, otherwise called noise, is in these data.

The effect of times on C-reactive protein was studied using traditional curvilinear regression. The traditional linear OLS regression provided a very poor fit for the C-reactive protein outcome data. Instead various curvilinear traditional regressions were assessed. A poor fit was shown for all of them. In fact, the best result was obtained by the S model with a p-value, however, of only 0,12 and an R square value of 0,106.

11.2.2 Summaries of Kernel Ridge Regressions

The best fit kernel density model was obtained by he Additive Chi square density model with an R square value of 0,131. This R square value was slightly better than that of the above traditional curvilinear regression S model with an p-value of 0,12 and an R square value of 0,106. This would mean that by the S model 10,6% of the certainty about the outcome was provided, while by the Additive chi square kernel density model 13,1% of the certainty about the outcome was provided.

11.3 Autoregression Analysis

For a proper assessment of seasonality, information of a second year of observation is needed, as well as information not only of, e.g., the months of January and July, but also of adjacent months. In order to unequivocally demonstrate seasonality, all of this information included in a single test is provided by autocorrelation.

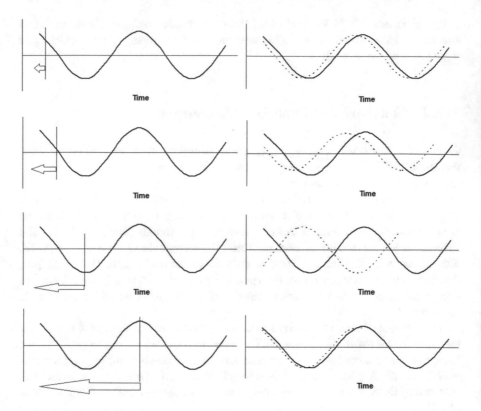

The above graph gives a simulated seasonal pattern of C-reactive protein levels in a healthy subject. Lag curves (dotted) are partial copies of the datacurve moved to the left as indicated by the arrows.

First-row graphs:the datacurve and the lag curve have largely simultaneous positive and negative departures from the mean, and, thus, have a strong positive correlation with one another (correlation coefficient ≈ +0.6).

Second-row graphs: this lag curve has little correlation with the datacurve anymore (correlation coefficient ≈ 0.0).

Third-row graphs: this lag curve has a strong negative correlation with the datacurve (correlation coefficient ≈ −1.0).

Fourth-row graphs: this lag curve has a strong positive correlation with the datacurve (correlation coefficient ≈ +1.0).

11.4 Data Example

Start by downloading the data file entitled "seasonality" in your computer mounted with SPSS statistical software 2022 version 28 0.1.0. For example, the first 10 months data are given underneath.

Average C-reactive protein in group of healthy subjects (mg/l)	Month
1,98	1
1,97	2
1,83	3
1,75	4
1,59	5
1,54	6
1,48	7
1,54	8
1,59	9
1,87	10

The standard errors of the averages were small, and, as compared to the autocorrelations computed, negligible, and were not taken into account.

11.5 Assessing Seasonality with Autocorrelations

Command:

Analyze....Forecasting....Autocorrelations....Move: monthly mean c-reactive protein levels into Variable Box....mark Autocorrelations....mark partial Autocorrelations.... click OK.

The output sheets are show the underneath graph.

VAR00008

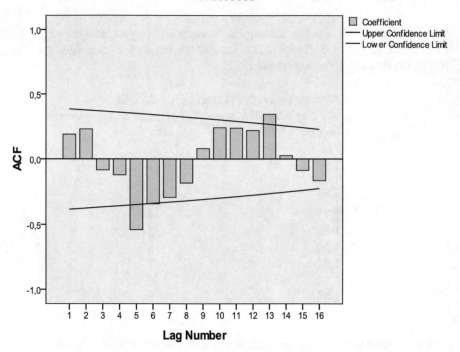

The above graph of monthly autocorrelation coefficients with their 95% confidence intervals is given by SPSS, and it shows that the magnitude of the monthly autocorrelations changes sinusoidally. The significant positive autocorrelations at the month no.13 (correlation coefficients of 0,42 (SE 0,14, t-value 3,0, p < 0,01)) further supports seasonality, and so does the pattern of partial autocorrelation coefficients (not shown): it gradually falls, and a partial autocorrelation coefficient of zero is observed 1 month after month 13. The strength of the seasonality is assessed using the magnitude of $r^2 = 0,42^2 = 0,18$. This would mean that the lag curve predicts the datacurve by only 18%, and, thus, that 82% is unexplained. And so, the seasonality may be statistically significant, but it is pretty weak, and a lot of unexplained variability, otherwise called noise, is in these data.

11.6   Assessing Seasonality with Curvilinear Regressions

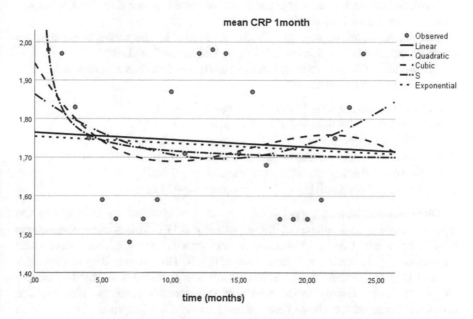

The effect of times on C-reactive protein was studied using traditional curvilinear regression. The traditional linear OLS regression provided a very poor fit for the C-reactive protein outcome data. Instead various curvilinear traditional regressions were assessed.

Open the datafile in SPSS version 2022, version 28 0.1.0.

Command: Menu....Analyze....Regression....Curvilinear....Dependent: C-reactive protein....Independent(s): Times....Model: mark Linear, quadratic Cubic, S, Exponential....click OK.

The above graph was in the output and it shows a poor fit for all of them. In fact, the best result was obtained by the S model with a p-value of only 0,12 and an R square value of 0,106.

Subsequently, kernel ridge regressions were performed.

11.7   Assessing Seasonality with Kernel Ridge Regressions

Command:

Menu....Analyze....Kernel Ridge Regression....Dependent: C-reactive protein.... Independent(s): times....click Linear....click OK.

In the output tables an R-square of −10,3.. is given. With traditional linear regression R square values are between 0 and 1. How is it possible that with kernel ridge regression R square values can, obviously, be negative!!! The answer is, that with kernel ridge regression a negative R square value is possible for models where a datafit is worse than horizontal.

Alternative kernel density models can be assessed. The above commands are repeated, but, instead of "Linear", "Additive Chi square" is clicked.

The Additive Chi square kernel density model produced an r square of 0,131.

More models were assessed.

The Chi square kernel density model   r square = −1922
The Cosine kernel density model   r square = −0,162
The Laplacian kernel model   r square = −10,137
The polynomial kernel density model   r square = −1210
The RBF kernel density model   r square = −12,926
The S kernel density model   r square = −0,172

Obviously the best fit kernel density model was obtained by he Additive Chi square density model with an R square value of 0,131. This R square value was slightly better than that of the above traditional curvilinear regression S model with an p-value of 0,12 and an R square value of 0,106. This would mean that by the S model 10,6% of the certainty about the outcome was provided, while by the Additive chi square kernel density model 13,1% of the certainty about the outcome was provided. Some of the above kernel density models are graphically presented in the Chap. 5 entitled "Some Terminology".

11.8   Conclusions

11.8.1   Summaries of Traditional Regressions

The monthly autocorrelation coefficients with their 95% confidence intervals show that the magnitude of the monthly autocorrelations changes sinusoidally. The significant positive autocorrelations at the month no.13 (correlation coefficients of 0,42 (SE 0,14, t-value 3,0, p < 0,01)) further supports seasonality, and so does the pattern of partial autocorrelation coefficients (not shown): it gradually falls, and a partial autocorrelation coefficient of zero is observed 1 month after month 13. The strength of the seasonality is assessed using the magnitude of $r^2 = 0,42^2 = 0,18$. This would mean that the lag curve predicts the datacurve by only 18%, and, thus, that 82% is unexplained. And so, the seasonality may be statistically significant, but it is pretty weak, and a lot of unexplained variability, otherwise called noise, is in these data. The effect of times on C-reactive protein also was studied using traditional curvilinear regression. The traditional linear OLS regression provided a very poor fit for the C-reactive protein outcome data. Instead various curvilinear traditional regressions were assessed. A poor fit was shown for all of them. In fact, the best result was obtained by the S model with a p-value, however, of only 0,12 and an R square value of 0,106.

11.8.2 Summaries of Kernel Ridge Regressions

The best fit kernel density model was obtained by he Additive Chi square density model with an R square value of 0,131. This R square value was slightly better than that of the above traditional curvilinear regression S model with an p-value of 0,12 and an R square value of 0,106. This would mean that by the S model 10,6% of the certainty about the outcome was provided, while by the Additive chi square kernel density model 13,1% of the certainty about the outcome was provided.

In conclusion, kernel ridge regression, although slightly better fit than curvilinear regressions, was not parsimonious to autocorrelation analysis.

11.9 References

All of the chapters of the current edition start with a brief review of the traditional analytic method of the different regression methods prior to the review of the relevant kernel ridge regression method. For the purpose, generally, data examples are used from the recent edition "Regression Analyses in Clinical Research for Starters and 2nd Levelers 2nd Edition, Springer Heidelberg Germany 2021", by the same authors. For a better understanding of differences between traditional and kernel ridge regressions, readers may benefit from the study of this edition first.

To readers requesting still more background, theoretical and mathematical information of computations given, several textbooks complementary to the current production and written by the same authors are available: Statistics applied to clinical studies 5th edition, 2012, Machine learning in medicine a complete overview 2nd edition, 2020, SPSS for starters and 2nd levelers 2nd edition, 2015, Clinical data analysis on a pocket calculator 2nd edition, 2016, Understanding clinical data analysis from published research, 2016, all of them edited by Springer Heidelberg Germany.

Chapter 12
Effect of Different Dosages of Prednisone and Beta-Agonist on Peakflow, 78 Patients, Traditional Regressions vs Kernel Ridge Regression

Abstract The effect of different dosages of prednisone and beta-agonists on peakflow was measured in 78 patients. In the traditional linear regression an R Square value of 0.582 is obtained, and the linear effects of prednisone dosages are a statistically significant predictor of the peak expiratory flow, but, surprisingly, the beta agonists dosages are not. A weighted least squares analysis was subsequently performed, which is a way for adjusting heteroscedasticity. The output table now shows an R Square value of 0.716. It has risen from 0,582, and provides thus better statistical power. With kernel ridge regressions the best fit R Square value was obtained from the Laplacian kernel density model, with a magnitude of 0,709, which is almost as large as that of the weighted least square analysis. The advantage of the kernel density analysis is that it is easier and does not require the pretty complex procedure of selection of the weight variable and calculation of likelihoods for different powers.

Keywords Weighted least squares analysis · Heteroscedasticity · Laplacian kernel density model

12.1 Summary

As data example the effect of different dosages of prednisone and beta-agonists on peakflow was measured in 40 patients. The spread of the outcome data may be smaller with low dosage treatments, than it may with high dosage treatment. The effect of prednisone on peak expiratory flow was assumed to be more variable with increasing dosages. Can the spread in the data, therefore, be measured with more precision, if linear regression is replaced with weighted least squares procedure. Does kernel ridge regression provide a meaningful alternative to traditional and weighted least squares regressions.

Supplementary Information The online version contains supplementary material available at [https://doi.org/10.1007/978-3-031-10717-7_12].

12.1.1   Summaries of Traditional Regressions

In the traditional linear regression an R Square value of 0.582 is obtained, and the linear effects of prednisone dosages are a statistically significant predictor of the peak expiratory flow, but, surprisingly, the beta agonists dosages are not. A weighted least squares analysis was subsequently performed, which is a way for adjusting heteroscedasticity. The output table now shows an R Square value of 0.716. It has risen from 0,582, and provides thus better statistical power. The effects of the two medicine dosages on the peak expiratory flows are shown. The t-values of the medicine predictors have increased from 10,217 and 0,597 to 13,740 and 3174. The p-values correspondingly fell from 0.000 and 0.552 to respectively 0.000 and 0.002. Obviously, after adjustment for heteroscedasticity, the beta agonist became a significant independent determinant of peakflow.

12.1.2   Summaries of Kernel Ridge Regressions

The best fit R Square value was obtained from the Laplacian kernel density model, with a magnitude of 0,709, which is almost as large as that of the weighted least square analysis. The advantage of the kernel density analysis is that it is easier and does not require the pretty complex procedure of selection of the weight variable and calculation of likelihoods for different powers. We should add, that one kernel density model, the sigmoid model, produced a negative R Square value. Kernel ridge density model can some time be negative. How is it possible that with kernel ridge regression R square values can, obviously, be negative!!! The answer is, that with kernel ridge regression a negative R square value is possible for models where a datafit is worse than horizontal.

12.2   Introduction

The spread of the outcome data may be smaller with low dosage treatments, than it may with high dosage treatment. The effect of prednisone on peak expiratory flow was assumed to be more variable with increasing dosages. Can it, therefore, be measured with more precision, if linear regression is replaced with weighted least squares procedure. Does kernel ridge regression provide a meaningful alternative to traditional and weighted least squares regressions. The data example below will used as analytic example.

12.3   Data Example (Var = Variable)

Var 1	Var 2	Var 3	Var 4
1	29	1,40	174
2	15	2,00	113
3	38	0,00	281
4	26	1,00	127
5	47	1,00	267
6	28	0,20	172
7	20	2,00	118
8	47	0,40	383
9	39	0,40	97
10	43	1,60	304
11	16	0,40	85
12	35	1,80	182
13	47	2,00	140
14	35	2,00	64
15	38	0,20	153
16	40	0,40	216

Var 1 Patient no
Var 2 prednisone (mg/24h)
Var 3 peak flow (ml/min)
Var 4 beta agonist (mg/24h)

Only the first 16 patients are given, the entire data file is entitled "weightedleast squares2" and is in SpringerLink supplementary files. We will first make a graph of prednisone dosages and peak expiratory flows. Start with opening the data file in your computer mounted with SPSS statistical software 2022 version 28 0.1.0.

12.4   Regressions with Inconstant Variability

Command:

click Graphs....Legacy Dialogs....Scatter/Dot....click Simple Scatter....click Define....Y Axis enter peakflow....X Axis enter prednisone....click OK.
   The underneath graph is in the output.

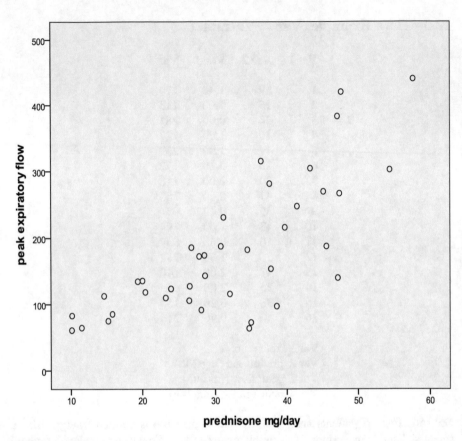

It shows, that the spread of the y-values is small with low dosages and gradually increases. We will, therefore, perform both a traditional, and a weighted least squares analysis of these data.

12.5   Traditional Linear Regression

Command:

Analyze....Regression....Linear....Dependent: enter peakflow....Independent: enter prednisone, betaagonist....click OK.

Model Summary[b]

Model	R	R Square	Adjusted R Square	Std. Error of the Estimate
1	,763[a]	,582	,571	65,304

a. Predictors: (Constant), beta agonist mg/24h, prednisone mg/day

b. Dependent Variable: peak expiratory flow

Coefficients[a]

Model		Unstandardized Coefficients		Standardized Coefficients		
		B	Std. Error	Beta	t	Sig.
1	(Constant)	-22,534	22,235		-1,013	,314
	prednisone mg/day	6,174	,604	,763	10,217	,000
	beta agonist mg/24h	6,744	11,299	,045	,597	,552

a. Dependent Variable: peak expiratory flow

In the above output sheets an R Square value of 0.582 is observed, and the linear effects of prednisone dosages are a statistically significant predictor of the peak expiratory flow, but, surprisingly, the beta agonists dosages are not.

We will subsequently perform a weighted least squares analysis, which is a way for adjusting heteroscedasticity.

12.6   Weighted Least Squares Regression

Command:

Analyze....Regression....Weight Estimation.... select: Dependent: enter peakflow .... Independent(s): enter prednisone, beta agonist....select prednisone also as Weight variable....Power range: enter 0 through 5 by 0.5....click Options....select Save best weights as new variable....click Continue....click OK.

In the output sheets it is observed, that the software has calculated likelihoods for different powers, and the best likelihood value is chosen for further analysis. When returning to the data file again a novel variable is added, the WGT_1 variable (the weights for the WLS analysis). The next step is, to perform again a linear regression, but now with the weight variable included.

Command:

Analyze....Regression....Linear.... select: Dependent: enter peakflow.... Independent(s): enter prednisone,
beta agonist....select the weights for the wls analysis (the GGT_1) variable as WLS Weight....click Save....select Unstandardized in Predicted Values....deselect Standardized in Residuals....click Continue....click OK.

Model Summary[b,c]

Model	R	R Square	Adjusted R Square	Std. Error of the Estimate
1	,846[a]	,716	,709	,125

a. Predictors: (Constant), beta agonist mg/24h, prednisone mg/day

b. Dependent Variable: peak expiratory flow

c. Weighted Least Squares Regression - Weighted by Weight for peakflow from WLS, MOD_6 PREDNISONE** -3,500

Coefficients[a,b]

Model		Unstandardized Coefficients		Standardized Coefficients		
		B	Std. Error	Beta	t	Sig.
1	(Constant)	5,029	7,544		,667	,507
	prednisone mg/day	5,064	,369	,880	13,740	,000
	beta agonist mg/24h	10,838	3,414	,203	3,174	,002

a. Dependent Variable: peak expiratory flow

b. Weighted Least Squares Regression - Weighted by Weight for peakflow from WLS, MOD_6 PREDNISONE** -3,500

The output table now shows an R Square value of 0.716. It has risen from 0,582, and provides thus better statistical power. The above lower table shows the effects of the two medicine dosages on the peak expiratory flows. The t-values of the medicine predictors have increased from 10,217 and 0,597 to 13,740 and 3174. The p-values correspondingly fell from 0.000 and 0.552 to respectively 0.000 and 0.002. Obviously, after adjustment for heteroscedasticity, the beta agonist became a significant independent determinant of peak flow.

12.7   Kernel Ridge Regression

Kernel ridge regressions were used as an alternative method. Eight different kernel density models were applied.

Command:

Menu....Analyze....Regression....Kernel Ridge Regression....Dependent: peak expiratory flow....Independent(s): prednisone, beta agonist....click Linear....click OK.

In the output sheets the underneath table shows a kernel ridge R Square value of 0,577, which is pretty good, but smaller than the traditional linear regression R Square of 0,582, and smaller than the R Square of 0,716 obtained through a special procedure for data with inconstant variability entitled "Weighted least square regression analysis".

Model Summary[a,b]

Kernel	Alpha	R Square
Linear	1,000	,577

a. Dependent Variable: peakflow

b. Model: prednisone, betaagonist

More kernel regression density models are possible. The same commands but with different kernel density models were subsequently analyzed for the purpose. The R Square results are underneath.

Model Summary[a,b]

Kernel	Alpha	R Square
Additive_chi2	1,000	,635

a. Dependent Variable: peakflow

b. Model: prednisone, betaagonist

Model Summary[a,b]

Kernel	Alpha	Gamma	R Square
Chi2	1,000	1,000	,619

a. Dependent Variable: peakflow

b. Model: prednisone, betaagonist

Model Summary[a,b]

Kernel	Alpha	R Square
Cosine	1,000	,017

a. Dependent Variable:
peakflow

b. Model: prednisone,
betaagonist

Model Summary[a,b]

Kernel	Alpha	Gamma	R Square
Laplacian	1,000	,500	,709

a. Dependent Variable: peakflow

b. Model: prednisone, betaagonist

Model Summary[a,b]

Kernel	Alpha	Gamma	Coef0	Degree	R Square
Polynomial	1,000	,500	1,000	3,000	,641

a. Dependent Variable: peakflow

b. Model: prednisone, betaagonist

Model Summary[a,b]

Kernel	Alpha	Gamma	R Square
RBF	1,000	,500	,599

a. Dependent Variable: peakflow

b. Model: prednisone, betaagonist

Model Summary[a,b]

Kernel	Alpha	Gamma	Coef0	R Square
Sigmoid	1,000	,500	1,000	-,001

a. Dependent Variable: peakflow

b. Model: prednisone, betaagonist

The best fit R Square value was obtained from the Laplacian kernel density model, with a magnitude of 0,709, which is almost as large as that of the weighted least square analysis. The advantage of the kernel density analysis is that it is easier and does not require the pretty complex procedure of selection of the weight variable and calculation of likelihoods for different powers. We should add, that one kernel density model, the sigmoid model, produced a negative R Square value. Kernel ridge density model can some time be negative. How is it possible that with kernel ridge regression R square values can, obviously, be negative!!! The answer is, that with kernel ridge regression a negative R square value is possible for models where a datafit is worse than horizontal. The above kernel ridge regressions make use of eight different kernel density distributions. All of them are graphically presented in the Chap. 5 entitled "Some Terminology".

12.8 Conclusion

The spread of the outcome data may be smaller with low dosage treatments, than it may with high dosage treatment. The effect of prednisone on peak expiratory flow was more variable with increasing dosages. The questions were the following. Can the spread in the data, therefore, be measured with more precision, if lincar regression is replaced with weighted least squares procedure. Does kernel ridge regression provide a meaningful alternative to traditional and weighted least squares regressions.

12.8.1 Summaries of Traditional Regressions

In the traditional linear regression an R Square value of 0,582 is obtained, and the linear effects of prednisone dosages are a statistically significant predictor of the peak expiratory flow, but, surprisingly, the beta agonists dosages are not. A weighted least squares analysis was subsequently performed, which is a way for adjusting heteroscedasticity. The output table now shows an R Square value of 0,716. It has risen from 0,582, and provides thus better statistical power. The above lower table shows the effects of the two medicine dosages on the peak expiratory flows. The t-values of the medicine predictors have increased from 10,217 and 0,597 to 13,740 and 3174. The p-values correspondingly fell from 0.000 and 0.552 to respectively 0,000 and 0,002. Obviously, after adjustment for heteroscedasticity, the beta agonist became a significant independent determinant of peak flow.

12.8.2 Summaries of Kernel Ridge Regressions

The best fit R Square value was obtained from the Laplacian kernel density model, with a magnitude of 0,709, which is almost as large as that of the weighted least square analysis. The advantage of the kernel density analysis is that it is easier and does not require the pretty complex procedure of selection of the weight variable and calculation of likelihoods for different powers. We should add, that one kernel density model, the sigmoid model, produced a negative R Square value. Kernel ridge density model can some time be negative. How is it possible that with kernel ridge regression R square values can, obviously, be negative!!! The answer is, that with kernel ridge regression a negative R square value is possible for models where a datafit is worse than horizontal. The above kernel ridge regressions make use of eight different kernel density distributions. All of them are graphically presented in the Chap. 5 entitled "Some Termminology" (see also the Chap. 5 entitled "Some Terminology").

12.9   References

All of the chapters of the current edition start with a brief review of the traditional regressions, and test it against kernel ridge regressions. For the purpose, generally, data examples are used from the recent edition "Regression Analyses in Clinical Research for Starters and 2nd Levelers 2nd Edition, Springer Heidelberg Germany 2021", by the same authors. For a better understanding of differences between traditional and kernel regressions, readers may benefit from the study of this edition first.

To readers requesting still more background, theoretical and mathematical information of computations given, several textbooks complementary to the current production and written by the same authors are available: Statistics applied to clinical studies 5th edition, 2012, Machine learning in medicine a complete overview, 2015, SPSS for starters and 2nd levelers 2nd edition, 2015, Clinical data analysis on a pocket calculator 2nd edition, 2016, Understanding clinical data analysis from published research, 2016, all of them edited by Springer Heidelberg Germany.

Chapter 13
Effect of Race, Age, and Gender on Physical Strength, 60 Patients, Traditional Regressions vs Kernel Ridge Regression

Abstract The effect of race, age, gender on physical strength was assessed in 60 patients. The variable race was traditionally assessed as a stepwise variable with four categories. and also in a restructured model as four binary variables (black yes or no, white yes or no, Asian yes or no, and hispanic yes or no). The R Square of the unrestructured model was 0,249 (24,9% certainty), that of the restructured model was 0,657 (65,7% certainty). With kernel ridge regressions the R Square of the best fit unrestructured kernel ridge model rose to 0,393. The R Square of the best fit restructured kernel density model rose to 0,693.

Keywords Restructured and unrestructured traditional regressions · Restructured and unrestructured kernel ridge regressions · Polynomial kernel density model

13.1 Summary

The effect of race, age, gender on physical strength was assessed in 60 patients. The variable race was assessed as a stepwise variable with four categories (1–4). and also in a slightly restructured model as four binary variables (black yes or no, white yes or no, Asian yes or no, and hispanic yes or no).

13.1.1 Summaries of Traditional Regressions

The R Square of the unrestructured model was 0,249, The F statistic and p-value of no difference from zero were 6183 and 0,001. The R Square of the restructured model was 0,657. The F statistic and p-value of no difference from zero were 20,728 and 0,000.

Supplementary Information The online version contains supplementary material available at [https://doi.org/10.1007/978-3-031-10717-7_13].

13.1.2   Summaries of Kernel Ridge Regressions

The R Square of the best fit unrestructured kernel ridge model was 0,393 using the Polynomial kernel density model, which was much better-fitted than that of the unrestructured traditional regression.

The R Square of the best fit restructured kernel density model was 0,693, again substantially better than that of the restructured traditional regression.

We should add that many kernel density models produced strong negative R Square values as a consequence of poor data fit of these models. The somewhat awkward result of negative R square values is observed with kernel ridge models where the data fit is worse than horizontal.

13.2   Introduction

The effect of race, age, gender on physical strength was assessed in 60 patients. The outcome of a linear regression is always a continuous variable. The predictor, however, can be continuous, binary or categorical. The equation of a linear regression is an incremental function in the form of a straight line, $y = a + bx$. If the predictor x increases, the outcome y will increase proportionally. With a binary predictor the x-values are often given the amount of 0 or 1, e.g. in therapeutic trials, a zero for the worse treatment, a one for the better treatment, and, so, this is an incremental function too. Similarly, a categorical predictor may be incremental, e.g. in therapeutic trials with incremental dosages of a drug. But, this is not always true, e.g. in a therapeutic trial with completely different compounds an incremental function may very well be lacking, and, thus, linear regression will be insignificant, and is an inappropriate model for assessment. Restructuring non-incremental categories into multiple binary variables is a solution for the problem.

13.3   Data Example

The datafile used in this chapter is entitled "restructuring categories", and is in SpringerLink supplementary files. For analysis it should be downloaded in your computer that is already mounted with SPSS statistical software 2022 version 28 0.1.0.

The first 10 patients of the 60 patient datafile is in the underneath table.

Variables

patient number	physical strength	race	age	gender
1	70,00	1,00	35,00	1,00
2	77,00	1,00	55,00	0,00
3	66,00	1,00	70,00	1,00
4	59,00	1,00	55,00	0,00
5	71,00	1,00	45,00	1,00
6	72,00	1,00	47,00	1,00
7	45,00	1,00	75,00	0,00
8	85,00	1,00	83,00	1,00
9	70,00	1,00	35,00	1,00
10	77,00	1,00	49,00	1,00

13.4 Restructuring Traditional Multiple Variables Regression

In a study with a categorical predictor like races, the race-values 1–4 have no incremental function, and, therefore, linear regression is not the best sensitive approach for assessing their effect on any outcome. Instead, restructuring the data for categorical predictors does the job. As an example, in a study the scientific question was: does race have an effect on physical strength. The variable race has a categorical rather than linear pattern. The effects on physical strength (scores 0–100) were assessed in 60 subjects of different races (hispanics (1), blacks (2), Asians (3), and whites (4)), ages (years), and genders (0 = female, 1 = male). The first 10 patients are in the table underneath.

patient number	physical strength	race	age	gender
1	70,00	1,00	35,00	1,00
2	77,00	1,00	55,00	0,00
3	66,00	1,00	70,00	1,00
4	59,00	1,00	55,00	0,00
5	71,00	1,00	45,00	1,00
6	72,00	1,00	47,00	1,00
7	45,00	1,00	75,00	0,00
8	85,00	1,00	83,00	1,00
9	70,00	1,00	35,00	1,00
10	77,00	1,00	49,00	1,00

The entire data file is in SpringerLink supplementary files, and is entitled "restructuring categories". It is previously used by the authors in SPSS for starters and 2nd levelers, Chap. 8, Springer Heidelberg Germany, 2016. Start by opening the data file in your computer with SPSS installed.

Command:

click race....click Edit....click Copy....click a new "var"....click Paste....highlight the values 2–4....delete and replace with 0,00 values....perform the same procedure subsequently for the other races.

patient number	physical strength	race	age	gender	race 1 hispanics	race 2 blacks	race 3 asians	race 4 whites
1	70,00	1,00	35,00	1,00	1,00	0,00	0,00	0,00
2	77,00	1,00	55,00	0,00	1,00	0,00	0,00	0,00
3	66,00	1,00	70,00	1,00	1,00	0,00	0,00	0,00
4	59,00	1,00	55,00	0,00	1,00	0,00	0,00	0,00
5	71,00	1,00	45,00	1,00	1,00	0,00	0,00	0,00
6	72,00	1,00	47,00	1,00	1,00	0,00	0,00	0,00
7	45,00	1,00	75,00	0,00	1,00	0,00	0,00	0,00
8	85,00	1,00	83,00	1,00	1,00	0,00	0,00	0,00
9	70,00	1,00	35,00	1,00	1,00	0,00	0,00	0,00
10	77,00	1,00	49,00	1,00	1,00	0,00	0,00	0,00

The result is shown above.

13.5   Unrestructured Traditional Regression

For the analysis we will use multiple linear regression. First a traditional linear regression analysis will be performed.

Command:

Analyze....Regression....Linear....Dependent: physical strength score.... Independent: race, age, gender....OK.

The upper table show an R Square value of 0,249, i.e., The predictors together determine the outcome strength by 24,9%. And so, 75,1% is unexplained. A percentage of determination under 25% indicates a very poor predictor model. Between 25 and 50% it means a reasonable, and over 50% it means a strong predictor model. The coefficients table shows, that age and gender are significant predictors but race is not.

Model Summary

Model	R	R Square	Adjusted R Square	Std. Error of the Estimate
1	,499ᵃ	,249	,209	12,11752

a. Predictors: (Constant), gender, race, age

ANOVAᵇ

Model		Sum of Squares	df	Mean Square	F	Sig.
1	Regression	2723,468	3	907,823	6,183	,001ᵃ
	Residual	8222,716	56	146,834		
	Total	10946,183	59			

a. Predictors: (Constant), gender, race, age
b. Dependent Variable: strengthscore

Coefficients^a

Model		Unstandardized Coefficients		Standardized Coefficients	t	Sig.
		B	Std. Error	Beta		
1	(Constant)	79,528	8,657		9,186	,000
	race	,511	1,454	,042	,351	,727
	age	-,242	,117	-,260	-2,071	,043
	gender	9,575	3,417	,349	2,802	,007

a. Dependent Variable: strengthscore

The variable race was analyzed as a stepwise rising function from category 1–4. and the linear regression model assumes, that the outcome variable will rise (or fall) simultaneously and linearly, but this needs not be necessarily so. And, therefore, this analysis may be somewhat flawed for these data. Next, a categorical analysis will be performed using a restructured datafile. The above commands are given once more, but now the independent variables are entered slightly differently.

13.6 Restructured Traditional Regression

Command:

Analyze....Regression....Linear....Dependent: physical strength score.... Independent: race 2, race 3, race 4, age, gender....click OK.

Model Summary

Model	R	R Square	Adjusted R Square	Std. Error of the Estimate
1	,811^a	,657	,626	8,33300

a. Predictors: (Constant), gender, race4, race2, age, race3

ANOVA[b]

Model		Sum of Squares	df	Mean Square	F	Sig.
1	Regression	7196,486	5	1439,297	20,728	,000[a]
	Residual	3749,698	54	69,439		
	Total	10946,183	59			

a. Predictors: (Constant), gender, race4, race2, age, race3
b. Dependent Variable: strengthscore

Coefficients[a]

Model		Unstandardized Coefficients		Standardized Coefficients	t	Sig.
		B	Std. Error	Beta		
1	(Constant)	72,650	5,528		13,143	,000
	race2	17,424	3,074	,559	5,668	,000
	race3	-6,286	3,141	-,202	-2,001	,050
	race4	9,661	3,166	,310	3,051	,004
	age	-,140	,081	-,150	-1,716	,092
	gender	5,893	2,403	,215	2,452	,017

a. Dependent Variable: strengthscore

The above tables show that the multiple binary variables model produces an R Square of 0,657, which is much better than the traditional multiple variables linear regression with R Square only 0,249.

The tables also show that race 2–4 are significant predictors of physical strength. The results can be interpreted as follows.

The underneath regression equation is used:

$$y = a + b_1 x_1 + b_2 x_2 + b_3 x_3 + b_4 x_4 + b_5 x_5$$

a = intercept
b_1 = regression coefficient for blacks (0 = no, 1 = yes),
b_2 = Asians
b_3 = whites
b_4 = age
b_5 = gender

If an individual is hispanic (race 1), then x_1, x_2, and x_3 will turn into 0, and the regression equation turn into $y = a + b_4x_4 + b_5x_5.$.

In summary:

if hispanic, $y = a + b_4x_4 + b_5x_5.$
if black, $y = a + b_1 + b_4x_4 + b_5x_5.$
if Asian, $y = a + b_2 + b_4x_4 + b_5x_5.$
if white, $y = a + b_3 + b_4x_4 + b_5x_5.$

So, e.g., the best predicted physical strength score of a white male of 25 years of age would equal.

$y = 72.65 + 9.66 - 0.14^* 25 + 5.89^*1 = 84.7$ (on a linear scale from 0 to 100),
   ($^* =$ sign of multiplication).

Compared to the presence of the hispanic race, the black and white races are significant positive predictors of physical strength ($p = 0,0001$ and $0,004$ respectively), the Asi.an race is a significant negative predictor ($p = 0,050$). All of these results are adjusted for age and gender, at least if we used $p = 0.,10$ as criterion for statistical significance.

The restructured data were first analyzed in a multiple variables OLS (ordinary least squares) linear regression as shown above. The races 2, 3 and 4 were different from race 1 at $p = 0,000, 0,050, 0,004$.

13.7   Unrestructured Kernel Ridge Regressions with Race as Categorical Predictor Variable

Your data must be downloaded from SpringerLink supplementary files into your computer installed with the 2022 version of SPSS statistical software 28.0.1.0. Start by opening the datafile.

Command:

Menu....Analyze....Regression....Kernel   Ridge   Regression....Dependent: strengthscore....Independent(s): race, age, gender....click Linear....click OK.

Model Summary[a,b]

Kernel	Alpha	R Square
Linear	1,000	-,886

a. Dependent Variable: strengthscore

b. Model: race, age, gender

A strong negative kernel ridge R Square of −0,888 is observed. More kernel density models are applied. Similar commands can be given.

Model Summary[a,b]

Kernel	Alpha	R Square
Additive_chi2	1,000	,285

a. Dependent Variable: strengthscore

b. Model: race, age, gender

Model Summary[a,b]

Kernel	Alpha	Gamma	R Square
Chi2	1,000	1,000	-,113

a. Dependent Variable: strengthscore

b. Model: race, age, gender

Model Summary[a,b]

Kernel	Alpha	R Square
Cosine	1,000	,001

a. Dependent Variable: strengthscore

b. Model: race, age, gender

Model Summary[a,b]

Kernel	Alpha	Gamma	R Square
Laplacian	1,000	,333	-,527

a. Dependent Variable: strengthscore

b. Model: race, age, gender

Model Summary[a,b]

Kernel	Alpha	Gamma	Coef0	Degree	R Square
Polynomial	1,000	,333	1,000	3,000	,393

a. Dependent Variable: strengthscore

b. Model: race, age, gender

Model Summary[a,b]

Kernel	Alpha	Gamma	R Square
RBF	1,000	,333	-1,794

a. Dependent Variable: strengthscore

b. Model: race, age, gender

Model Summary[a,b]

Kernel	Alpha	Gamma	Coef0	R Square
Sigmoid	1,000	,333	1,000	-,008

a. Dependent Variable: strengthscore

b. Model: race, age, gender

The polynomial density model of the kernel ridge regression obviously produced an R Square much better than the traditional linear regression with R Square values of 0,393 vs 0,249. We should add, that many models produced the awkward result of negative R square values. We should emphasize that a negative result here is not erroneous. See also the Chap. 5, entitled "Some Terminology". The above kernel ridge regressions make use of eight different kernel density models. All of them are graphically presented in the Chap. 5 entitled "Some Terminology".

13.8   Restructured Kernel Ridge Regressions

The restructured version of the dataset with multiple binary race predictors instead of one was previously applied and provided better fit results. We will now provide this restructured version also for kernel regressions. Again, your data must be downloaded from SpringerLink supplementary files into your computer installed with the 2022 version of SPSS statistical software 28.0.1.0. Start by opening the datafile.

Command:

Menu....Analyze....Regression....Kernel   Ridge   Regression....Dependent:
strengthscore....Independent(s): race2, race3, race4, age, gender....click Linear....
click OK.

Model Summary[a,b]

Kernel	Alpha	R Square
Linear	1,000	-,462

a. Dependent Variable:
   strengthscore

b. Model: race2, race3,
   race4, age, gender

The kernel ridge regression output with a linear density model produced a very
weak results with a very negative R Square of −0,462. More models were applied,
and similar commands were given for the purpose.

Model Summary[a,b]

Kernel	Alpha	R Square
Additive_chi2	1,000	,658

a. Dependent Variable:
   strengthscore

b. Model: race2, race3, race4,
   age, gender

Model Summary[a,b]

Kernel	Alpha	Gamma	R Square
Chi2	1,000	1,000	-,356

a. Dependent Variable: strengthscore

b. Model: race2, race3, race4, age,
   gender

Model Summary[a,b]

Kernel	Alpha	R Square
Cosine	1,000	,002

a. Dependent Variable: strengthscore

b. Model: race2, race3, race4, age, gender

Model Summary[a,b]

Kernel	Alpha	Gamma	R Square
Laplacian	1,000	,200	,093

a. Dependent Variable: strengthscore

b. Model: race2, race3, race4, age, gender

Model Summary[a,b]

Kernel	Alpha	Gamma	Coef0	Degree	R Square
Polynomial	1,000	,200	1,000	3,000	,693

a. Dependent Variable: strengthscore

b. Model: race2, race3, race4, age, gender

Model Summary[a,b]

Kernel	Alpha	Gamma	R Square
RBF	1,000	,200	-,920

a. Dependent Variable: strengthscore

b. Model: race2, race3, race4, age, gender

Model Summary[a,b]

Kernel	Alpha	Gamma	Coef0	R Square
Sigmoid	1,000	,200	1,000	-,008

a. Dependent Variable: strengthscore

b. Model: race2, race3, race4, age, gender

Particularly, the Additive CHI2 kernel density model produced an R Square value of no less than 0, 658, and the polynomial kernel density model produced an R Square even better 0,693. This was better than that of the restructured binary predictive variables model with an R Square of 0,657. And so, this data sample is an example of a model where kernel ridge regression can be parsimonious to traditional linear regression, and also where the kernel ridge regressions of the restructured version were several times parsimonious to the non-restructured version. The above kernel ridge regressions make use of eight different kernel density models. All of them are graphically presented in the Chap. 5 entitled "Some Terminology".

13.9 Conclusion

In this chapter the effect of race, age, gender on physical strength was assessed in 60 patients. The variable race was assessed as a stepwise variable with four categories (1–4). and also in a slightly restructured model as four binary variables (black yes or no, white yes or no, Asian yes or no, and hispanic yes or no).

13.9.1 Summaries of Traditional Regressions

The R Square of the unrestructured model was 0,249, The F statistic and p-value of no difference from zero were 6183 and 0,001. The R Square of the restructured model was 0,657. The F statistic and p-value of no difference from zero were 20,728 and 0,000.

13.9.2 Summaries of Kernel Ridge Regressions

The R Square of the best fit unrestructured kernel ridge model was 0,393 using the Polynomial kernel density model, which was much better-fitted than that of the unrestructured traditional regression.

The R Square of the best fit restructured kernel density model was 0,693, again substantially better than that of the restructured traditional regression.

We should add that many kernel density models produced strong negative R Square values as a consequence of poor datafit of these models.

13.10 References

All of the chapters of the current edition start with a brief review of the traditional analytic method of the different regression methods prior to the review of the relevant kernel ridge regression method. For the purpose, generally, data examples are used from the recent edition "Regression Analyses in Clinical Research for

Starters and 2nd Levelers 2nd Edition, Springer Heidelberg Germany 2021", by the same authors. For a better understanding of differences between traditional and kernel regressions, readers may benefit from the study of this edition first.

To readers requesting still more background, theoretical and mathematical information of computations given, several textbooks complementary to the current production and written by the same authors are available: Statistics applied to clinical studies 5th edition, 2012, Machine learning in medicine a complete overview, 2015, SPSS for starters and 2nd levelers 2nd edition, 2015, Clinical data analysis on a pocket calculator 2nd edition, 2016, Understanding clinical data analysis from published research, 2016, all of them edited by Springer Heidelberg Germany.

Chapter 14
Effect of Treatment, Age, Gender, and Co-morbidity on Hours of Sleep, 20 Patients, Traditional Regression vs Kernel Ridge Regression

Abstract In a 20 patient parallel group trial the effect of treatment modality and covariates on hours of sleep was studied. The overall R Square value was 0,970, meaning that the outcome is predicted by the predictors by 97%. Multicollinearity was, however, in the data. The R Square of the kernel ridge regression using a Linear kernel density model was 0,963, which demonstrates a very good data fit virtually similar to that of the traditional multiple variables regression of 0,970. But kernel ridge regression is better, because it is adjusted for multicollinearity.

Keywords Parallel group trial · Multicollinearity · Linear kernel density model

14.1 Summary

In a 20 patient parallel group trial the effect of treatment modality (sleeping pill or placebo) and other covariates on hours of sleep was studied.

14.1.1 Summaries of Traditional Linear Regressions

The overall R Square value was 0,970, meaning that the outcome is predicted by the predictors by 97%. This 97% predictive certainty was significantly different from a 0% predictive certainty at $p < 0.000$. However, on clinical grounds the presence of multicollinearity was suspected. And, therefore, its presence was assessed with one-by-one linear regressions: if $R > 0,85$, then its presence is confirmed, and the model is no longer valid. One of the two variables responsible must be removed.

Supplementary Information The online version contains supplementary material available at [https://doi.org/10.1007/978-3-031-10717-7_14].

14.1.2 Summaries of Kernel Ridge Regressions

The R Square of the Linear kernel density model was 0,963 which demonstrates a very good data fit virtually similar to that of the traditional multiple variables regression of 0,970. But kernel ridge regression is better, because it is adjusted for multicollinearity. Also in the output is a graph of a scatterplot of predicted values versus treatment effects. A very nice linear data pattern is observed in the underneath graph.

14.2   Introduction

In a 20 patients parallel group trial the effect of treatment modality (sleeping pill or placebo) and other covariates on hours of sleep was studied. In a multiple variables linear regression analysis the predictors treatment modality, age, gender (male/female), and co-morbidity were assessed. Treatment modality and age were significant predictors at p = 0,010 and p = 0,000. The overall R-Square of the traditional linear regression was 0,970. However, this analysis model suffered from multicollinearity. Ridge Regression is a technique for analyzing multiple regression data that suffer from multicollinearity. **By adding a degree of bias to the regression estimates**, ridge regression reduces the standard errors. It is hoped that the net effect will be to give estimates that are more accurate without loss of sensitivity of testing.

14.3   Data Example

Primary scientific question: is the sleeping pill more efficaceous than the placebo on hours of sleep. Secondary question: do age, gender, and co-morbidity have additional predictive properties?

20 patients
Variable (Var)

1	2	3	4	5
0,00	6,00	65,00	0,00	1,00
0,00	7,10	75,00	0,00	1,00
0,00	8,10	86,00	0,00	0,00
0,00	7,50	74,00	0,00	0,00
0,00	6,40	64,00	0,00	1,00
0,00	7,90	75,00	1,00	1,00
0,00	6,80	65,00	1,00	1,00
0,00	6,60	64,00	1,00	0,00
0,00	7,30	75,00	1,00	0,00
0,00	5,60	56,00	0,00	0,00
1,00	5,10	55,00	1,00	0,00
1,00	8,00	85,00	0,00	1,00
1,00	3,80	36,00	1,00	0,00
1,00	4,40	47,00	0,00	1,00
1,00	5,20	58,00	1,00	0,00
1,00	5,40	56,00	0,00	1,00
1,00	4,30	46,00	1,00	1,00
1,00	6,00	64,00	1,00	0,00
1,00	3,70	33,00	1,00	0,00
1,00	6,20	65,00	0,00	1,00

Var 1 = group 0 has placebo, group 1 has sleeping pill
Var 2 = hours of sleep (effect treatment)
Var 3 = age
Var 4 = gender
Var 5 = co-morbidity

14.4   Traditional Regression

The above datafile is in SpringerLink supplementary files, and is entitled "hours of sleep". It must be downloaded into your computer mounted with SPSS Statistical software 2022 version 28 0.1.0.

   Then command.

Command:

Analyze....Regression....Linear....Dependent: hours of sleep....Independent (s): group, age, male/female, comorbidity....click OK.

   The underneath tables are in the output sheets.

Model Summary

Model	R	R Square	Adjusted R Square	Std. Error of the Estimate
1	,985[a]	,970	,963	,26568

a. Predictors: (Constant), comorbidity, group, male/female, age

ANOVA[a]

Model		Sum of Squares	df	Mean Square	F	Sig.
1	Regression	34,763	4	8,691	123,128	,000[b]
	Residual	1,059	15	,071		
	Total	35,822	19			

a. Dependent Variable: effect treatment

b. Predictors: (Constant), comorbidity, group, male/female, age

Coefficients[a]

Model		Unstandardized Coefficients		Standardized Coefficients	t	Sig.
		B	Std. Error	Beta		
1	(Constant)	,727	,406		1,793	,093
	group	-,420	,143	-,157	-2,936	,010
	age	,087	,005	,912	16,283	,000
	male/female	,202	,138	,075	1,466	,163
	comorbidity	,075	,130	,028	,577	,573

a. Dependent Variable: effect treatment

The overall R Square value was no less than 0,970, meaning that the outcome is predicted by the predictors by 97%. The analysis of variance showed an F-value of 123,128, and a p-value of ,000. The 97% predictive certainty is significantly different from a 0% predictive certainty at p < 0,000.

However, on clinical grounds the presence of multicollinearity in the data was suspected. And, therefore, its presence was assessed with one-by-one linear regressions: if R > 0,85, then multicollinearity is in the data, and the model is no longer valid. One of the two variables responsible must be removed. For assessment the underneath commands are required.

Command:

Analyze....Correlate....Bivariate....Variables: enter group, effect treatment, age, male/female, comorbidity....mark Pearson.....click OK.

A one by one linear regression table with Pearson correlation coefficients is in the output.

Correlations

		group	effect treatment	age	male/female	comorbidity
group	Pearson Correlation	1	-,643**	-,549*	,200	,000
	Sig. (2-tailed)		,002	,012	,398	1,000
	N	20	20	20	20	20
effect treatment	Pearson Correlation	-,643**	1	,975**	-,299	,134
	Sig. (2-tailed)	,002		,000	,201	,572
	N	20	20	20	20	20
age	Pearson Correlation	-,549*	,975**	1	-,364	,150
	Sig. (2-tailed)	,012	,000		,115	,528
	N	20	20	20	20	20
male/female	Pearson Correlation	,200	-,299	-,364	1	-,400
	Sig. (2-tailed)	,398	,201	,115		,081
	N	20	20	20	20	20
comorbidity	Pearson Correlation	,000	,134	,150	-,400	1
	Sig. (2-tailed)	1,000	,572	,528	,081	
	N	20	20	20	20	20

**. Correlation is significant at the 0.01 level (2-tailed).
*. Correlation is significant at the 0.05 level (2-tailed).

Multicollinearity was obviously in the data, because age by effect treatment and, of course, effect treatment by age produced R values of 0,975. The model is no longer valid. One of the two variables responsible must be either removed or the traditional analysis model must be replaced with a ridge regression.

Ridge Regression is a technique for analyzing multiple regression data that suffer from multicollinearity. **By adding a degree of bias to the regression estimates**, ridge regression reduces the standard errors. It is hoped that the net effect will be to give estimates that are more accurate without loss of sensitivity of testing.

14.5 Kernel Ridge Regression

Command

Analyze....Regression....Kernel Ridge Regression....Dependent: effect treatment.... Independent(s): treatment modality (group), age, gender, comorbidity....click

Linear....click   Options....mark   Observed   vs.   Predicted....mark
Residuals vs. Predicted....click OK.

The result of the linear kernel density model in the underneath table.

Model Summary[a,b]

Kernel	Alpha	R Square
Linear	1,000	,963

a. Dependent Variable: treatment

b. Model: group, age, gender, comorbidity

The R Square of the kernel density model Linear is 0,963 which demonstrates a very good datafit virtually similar to that of the traditional multiple variables regression of 0,970. But kernel ridge regression is better, because it is adjusted for multicollinearity.

Also in the output is a graph of a scatterplot of predicted values versus treatment effects. A very nice linear data pattern is observed in the underneath graph.

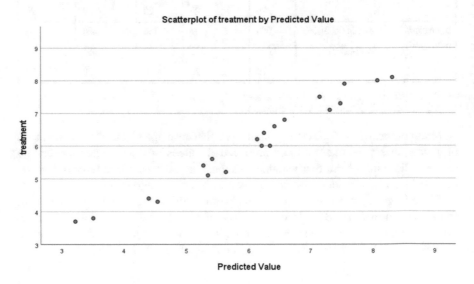

The Linear kernel ridge regression makes use of Gaussian distributions. Graphical presentations of some of them are in the Chap. 5, entitled "Some terminology".

14.6 Conclusion

In a 20 patients parallel group trial the effect of treatment modality (sleeping pill or placebo) and of covariates on hours of sleep was studied.

14.6.1 Summaries of Traditional Linear Regressions

The overall R Square value was 0,970, meaning that the outcome is predicted by the predictors by 97%. This 97% predictive certainty was significantly different from a 0% predictive certainty at $p < 0.000$. However, on clinical grounds the presence of multicollinearity was suspected. And, therefore, its presence was assessed with one-by-one linear regressions: if $R > 0,85$, then its presence is confirmed, and the model is no longer valid. One of the two variables responsible must be removed.

14.6.2 Summaries of Kernel Ridge Regressions

The R Square of the Linear kernel density model was 0,963 which demonstrates a very good datafit virtually similar to that of the traditional multiple variables regression of 0,970. But kernel ridge regression is better, because it is adjusted for multicollinearity. Also in the output is a graph of a scatterplot of predicted values versus treatment effects. A very nice linear data pattern is observed in the underneath graph.

14.7 References

All of the chapters of the current edition start with a brief review of the traditional analytic method of the different regression methods prior to the review of the relevant kernel ridge regression method. For the purpose, generally, data examples are used from the recent edition "Regression Analyses in Clinical Research for Starters and 2nd Levelers 2nd Edition, Springer Heidelberg Germany 2021", by the same authors. For a better understanding of differences between traditional and alternative regressions, readers may benefit from the study of this edition first.

To readers requesting still more background, theoretical and mathematical information of computations given, several textbooks complementary to the current production and written by the same authors are available: Statistics applied to clinical studies 5th edition, 2012, Machine learning in medicine a complete overview, 2015, SPSS for starters and 2nd levelers 2nd edition, 2015, Clinical data analysis on a pocket calculator 2nd edition, 2016, Understanding clinical data analysis from published research, 2016, all of them edited by Springer Heidelberg Germany.

Chapter 15
Effect of Counseling and Non-compliance on Monthly Stools, 35 Constipated Patients, Traditional Regressions vs Kernel Ridge Regression

Abstract The effect on monthly stools of counseling and non-compliance was assessed in 35 constipated patients. An output table showed, that counseling is a very significant predictor of therapeutic efficacy (stools) and non-compliance also a borderline significant predictor of stools at p = 0,098. The multiple variables regression produced an R Square of 0, 409. Next a one by one linear regression shows, that non-compliance is also a significant predictor of counseling at p = 0,024. Therefore, two stage least square (2SLS) method may be a better sensitive method. It works as follows. First, a simple regression analysis, with counseling as outcome and non-compliance as predictor, is performed. Then the outcome values of the regression equation are used as predictor of therapeutic efficacy. The R Square of the 2SLS procedure produced, however, an R Square of only 0,118. A better result is welcome.

With kernel ridge regressions better results were obtained than those of the traditional methodologies (traditional linear regression and two stage least squares, resp. R Squares 0,409 and 0,118) namely.

Additive_chi2 kernel density model R Square 0,472
Polynomial kernel density model R Square 0,570.

Keywords Two stage least squares · Additive chi2 kernel density model · Polynomial kernel density model

15.1 Summary

The effect on monthly stools of counseling and non-compliance was assessed in 35 constipated patients.

Supplementary Information The online version contains supplementary material available at [https://doi.org/10.1007/978-3-031-10717-7_15].

15.1.1 Summaries of Traditional Regressions

The above table shows, that counseling is a very significant predictor of therapeutic efficacy (stools) and non-compliance also a borderline significant predictor of stools at p = 0,098. The multiple variables regression produced an R Square of 0, 409. Next a one by one linear regression shows, that non-compliance is also a significant predictor of counseling at p = 0,024. Two stage least square (2SLS) method is available in SPSS. It works as follows. First, a simple regression analysis, with counseling as outcome and non-compliance as predictor, is performed. Then the outcome values of the regression equation are used as predictor of therapeutic efficacy. The R Square of the 2SLS procedure produced, however, an R Square of only 0,118. A better result is welcome.

15.1.2 Summaries of Kernel Ridge Regressions

The best fit R Square values of the various kernel ridge regression models were better than those of the traditional methodologies (traditional linear regression and two stage least squares, resp. R Squares 0,409 and 0,118):

Additive_chi2 kernel density model R Square 0,472
Polynomial kernel density model R Square 0, 570.

The scatterplot from the polynomial kernel density model of predicted values by therapeutic efficacies provided a nice linear data pattern, while the scatterplot of the residuals of these models provided a corresponding inversed linear pattern.

15.2 Introduction

The effect on monthly stools of counseling and non-compliance was assessed in 35 constipated patients. The two stage least squares (2SLS) method can also be applied, if heterogeneous variances due to interactions are expected. Heteroscedasticity is the phenomenon, that variability in a variable is unequal across the range of values of a second variable that predicts it. It is often due to error in the outcome derived from an independent variable, if another independent variable is interacting. As a data example, the effects of counseling and non-compliance (pills not used) on the efficacy of a novel drug may suffer from both heterogeneous variances and heteroscedasticity.

15.3   Data Example

In this chapter the effects of counseling and non-compliance (pills not used) on the efficacy of a novel laxative drug will be studied in 35 patients. The first 10 patients of the data file is below. The entire datafile is in SpringerFile supplementary files and is entitled "twostageleastsquares". Non-compl = non-compliance.

Variables (Var)

Var 1 (x)	var 2 (y)	var 3 (z)
Predictor 1	Outcome	Predictor 2
Explanatory	Therapeutic efficacy	Instrumental
non-compl	stools / month	counseling
1. 8	25	24
2. 13	30	30
3. 15	25	25
4. 14	31	35
5. 9	36	39
6. 10	33	30
7. 8	22	27
8. 5	18	14
9. 13	14	39
10. 15	30	42

The data file is in SpringerLink supplementary files and is entitled "twostageleastsquares". It was previously used by the authors in Machine learning in medicine a complete overview, Chap. 34, Springer Heidelberg Germany, 2015. Start by opening the file in your computer installed with SPSS statistical software 2022 version 28 0.1.0.

15.4   Traditional Regression Analysis

Command:

Analyze....Regression....Linear....Dependent: enter the eff (stool/month).... Independent(s): counseling and non-compliance....click OK.

The underneath tables are in the output.

Model Summary

Model	R	R Square	Adjusted R Square	Std. Error of the Estimate
1	,639[a]	,409	,372	8,049

a. Predictors: (Constant), counseling, non-compliance

Coefficients[a]

Model		Unstandardized Coefficients		Standardized Coefficients	t	Sig.
		B	Std. Error	Beta		
1	(Constant)	2,270	4,823		,471	,641
	counseling	1,876	,290	,721	6,469	,000
	non-compliance	,285	,167	,190	1,705	,098

a. Dependent Variable: ther eff

The above table shows, that counseling is a very significant predictor of therapeutic efficacy (stools) and non-compliance also a borderline significant predictor of stools at $p = 0,098$.

Next a one by one linear regression shows, that non-compliance is also a significant predictor of counseling at $p = 0,024$.

Command:

Analyze....Regression....Linear....Dependent: enter counseling....
    Independent(s): non-compliance....click OK.

Coefficients[a]

Model		Unstandardized Coefficients		Standardized Coefficients	t	Sig.
		B	Std. Error	Beta		
1	(Constant)	4,228	2,800		1,510	,141
	non-compliance	,220	,093	,382	2,373	,024

a. Dependent Variable: counseling

This would mean, that non-compliance works two ways: it predicts therapeutic efficacy *directly* and *indirectly* through counseling. However, the indirect way is not taken into account in the usual one step linear regression.

15.5   Two Stage Least Squares (2SLS)

Two stage least square (2SLS) method is possible, and is available in SPSS. It works as follows. First, a simple regression analysis, with counseling as outcome and non-compliance as predictor, is performed. Then the outcome values of the regression equation are used as predictor of therapeutic efficacy.

Command:

Analyze....Regression....2 Stage Least Squares....Dependent: therapeutic efficacy....Explanatory: non-compliance....Instrumental: counseling .... click OK.

Model Description

		Type of Variable
Equation 1	VAR00001	dependent
	VAR00003	predictor
	VAR00002	instrumental

MOD_1

Model Summary

Equation 1	Multiple R	,344
	R Square	,118
	Adjusted R Square	,092
	Std. Error of the Estimate	17,829

Coefficients

		Unstandardized Coefficients		Beta	t	Sig.
		B	Std. Error			
Equation 1	(Constant)	-61,095	37,210		-1,642	,110
	VAR00003	3,113	1,256	2,078	2,478	,019

The above tables show the results of the 2SLS method. As expected the final p-value is smaller than the simple linear regression p-value of the effect of non-compliance on therapeutic efficacy (p-value = 0,019 instead of 0,098. However, this is not accompanied by a larger R square value (R Square 0,118 instead of 0,409). Although the R Square value is at a reasonable predictive level, a better R Square is welcome. The kernel ridge methodology will be applied next.

15.6   Kernel Ridge Regression

Command in your computer mounted with SPSS 2022 version 18.0.1.0:

Command:

Menu....Analyze....Regression....Kernel Ridge Regression....Dependent: y....Independent(s): x, z....click Linear....click Options: mark Observed vs Predicted....click Continue....click OK.

Model Summary[a,b]

Kernel	Alpha	R Square
Linear	1,000	,385

a. Dependent Variable: y
b. Model: x, z

The above table gives an R Square of 0,385 (38,5% certainty of prediction), which is not bad, but worse than the R Square from the traditional linear regression (0,409). A better R Square is welcome. Additional kernel density models were assessed with similar commands.

Model Summary[a,b]

Kernel	Alpha	R Square
Additive_chi2	1,000	,472

a. Dependent Variable: y
b. Model: x, z

Model Summary[a,b]

Kernel	Alpha	Gamma	R Square
Chi2	1,000	1,000	,298

a. Dependent Variable: y

b. Model: x, z

Model Summary[a,b]

Kernel	Alpha	R Square
Cosine	1,000	,123

a. Dependent Variable: y

b. Model: x, z

Model Summary[a,b]

Kernel	Alpha	Gamma	R Square
Laplacian	1,000	,500	-,189

a. Dependent Variable: y

b. Model: x, z

Model Summary[a,b]

Kernel	Alpha	Gamma	Coef0	Degree	R Square
Polynomial	1,000	,500	1,000	3,000	,570

a. Dependent Variable: y

b. Model: x, z

Model Summary[a,b]

Kernel	Alpha	Gamma	R Square
RBF	1,000	,500	-,754

a. Dependent Variable: y

b. Model: x, z

Model Summary[a,b]

Kernel	Alpha	Gamma	Coef0	R Square
Sigmoid	1,000	,500	1,000	-,007

a. Dependent Variable: y

b. Model: x, z

The best fit R Square values of the various kernel ridge regression models were better than the those of the traditional methodologies (traditional linear regression and two stage least squares, resp. R Square 0,409 and 0,118):

Additive_chi2 kernel density model R Square 0,472
Polynomial kernel density model R Square 0, 570.

Note: Three of the above kernel density models produced very poor and even negative R Square values. The explanation in in Chap. 5 entitled "Some Terminology".

Also in the output were the underneath scatterplots of predicted values by therapeutic efficacies as well as scatterplots of predicted residual values. These scatterplots were from the polynomial kernel density model and provided nice linear data patterns, while the scatterplot of the residuals of this model provided a corresponding inversed linear pattern.

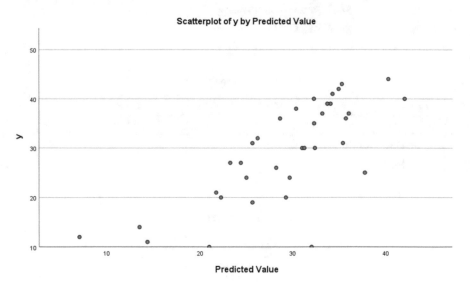

Scatterplot of y by Predicted Value

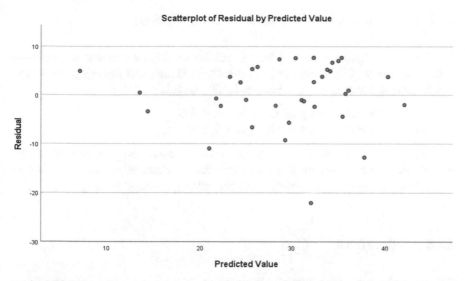

We should add, that the above analyses use eight different kernel density models. Graphical presentations of all of them are in the Chap. 5, entitled "Some Terminology".

15.7   Conclusion

In 35 constipated patients the effect on monthly stools of counseling and non-compliance was assessed.

15.7.1   *Summaries of Traditional Regressions*

The above table shows, that counseling is a very significant predictor of therapeutic efficacy (stools) and non-compliance also a borderline significant predictor of stools at $p = 0,098$. The multiple variables regression produced an R Square of 0, 409. Next a one by one linear regression shows, that non-compliance is also a significant predictor of counseling at $p = 0,024$. Two stage least square (2SLS) method is available in SPSS. It works as follows. First, a simple regression analysis, with counseling as outcome and non-compliance as predictor, is performed. Then the outcome values of the regression equation are used as predictor of therapeutic efficacy. The R Square of the 2SLS procedure produced, however, an R Square of only 0,118. A better result is welcome.

15.7.2 Summaries of Kernel Ridge Regressions

The best fit R Square values of the various kernel ridge regression models were better than those of the traditional methodologies (traditional linear regression and two stage least squares, resp. R Squares 0,409 and 0,118):

Additive_chi2 kernel density model R Square 0,472
Polynomial kernel density model R Square 0, 570.

Scatterplots from the polynomial kernel density model of predicted values by therapeutic efficacies provided a nice linear data pattern, while the scatterplot of the residuals of this model provided a corresponding inversed linear pattern.

15.8 References

All of the chapters of the current edition start with a brief review of the traditional analytic method of the different regression methods prior to the review of the relevant kernel ridge regression method. For the purpose, generally, data examples are used from the recent edition "Regression Analyses in Clinical Research for Starters and 2nd Levelers 2nd Edition, Springer Heidelberg Germany 2021", by the same authors. For a better understanding of differences between traditional and kernel regressions, readers may benefit from the study of this edition first.

To readers requesting still more background, theoretical and mathematical information of computations given, several textbooks complementary to the current production and written by the same authors are available: Statistics applied to clinical studies 5th edition, 2012, Machine learning in medicine a complete overview, 2015, SPSS for starters and 2nd levelers 2nd edition, 2015, Clinical data analysis on a pocket calculator 2nd edition, 2016, Understanding clinical data analysis from published research, 2016, all of them edited by Springer Heidelberg Germany.

Chapter 16
Effect of Treatment Modality, Counseling, and Satisfaction with Doctor on Quality of Life, 450 Patients, Traditional Regressions vs Kernel Ridge Regression

Abstract The effect of three predictors on health related quality of life scores was assessed with traditional and kernel ridge regressions. The R Square value of the traditional linear regression was 0,105, which indicates 10,5% certainty to predict the outcome qol score. This is a poor result. R Square values under 0,25 are assumed to have very poor predictive potential. The usually applied Cox and Snell R Square of the multinomial regression was larger than the R Square of a traditional linear regression (0,191 vs 0,105). The overall Cox and Snell pseudo R Square of the ordinal regression model was only 0,117, smaller than that of the multinomial regression (0,191).

With kernel ridge density models much better results were obtained.

the Chi2 kernel density model (R Square 0,241),
the Laplacian kernel density model (R Square 0,262),
the Polynomial kernel density model (R Square 0,205),
the RBF kernel density model (R Square 0,237).

Keywords Cox and Snell pseudo R squares · Multinomial regressions · Kernel ridge regressions

16.1 Summary

With traditional multivariables regression the outcome should be continuous, and in the example of the current Chap. the outcome qol score is measured in only 5 steps. Nonetheless the SPSS program provides a result of such an analysis. However, because of model's poor fit the R square is small, only 0,105. Better datafit should be obtained by several alternative analysis assessments. As data example the effect of three predictors on health related quality of life was assessed with traditional and kernel ridge regressions.

Supplementary Information The online version contains supplementary material available at [https://doi.org/10.1007/978-3-031-10717-7_16].

16.1.1   Summaries of Traditional Regressions

The R Square value of the traditional linear regression was 0,105, which indicates 10,5% certainty to predict the outcome qol score. This is a poor result. R Square values under 0,25 are assumed to have very poor predictive potential. The usually applied Cox and Snell R Square of the multinomial regression was larger than the R Square computed by the traditional linear regression (0,191 vs 0,105). The overall Cox and Snell pseudo R Square of the ordinal regression model was only 0,117, smaller than that of the multinomial regression (0,191).

16.1.2   Summaries of Kernel Density Regressions

Kernel ridge density models show, that, particularly

the Chi2 kernel density model (R Square 0,241),
the Laplacian kernel density model (R Square 0,262),
the Polynomial kernel density model (R Square 0,205),
the RBF kernel density model (R Square 0,237)

performed (much) better than did the above traditional regression methodologies. A kernel density R Square over 0,250 like the one with the Laplacian model is assumed to be a reasonable result in terms of predictive performance.

16.2   Introduction

With traditional multivariables regression the outcome should be continuous, and in the example of the current Chap. the outcome qol score is measured in only 5 steps. Nonetheless the SPSS program provides a result of such an analysis. However, because of poor fit the R square is small, only 0,105. Better datafit should be obtained by alternative analysis models.

Various closely related methodologies are available for analyzing categorical predictor variables.

1. *Multinomial regression* is fine for predictors without a stepping function,
2. *Ordinal regression* should be used if the predictor is stepwise.

These methodologies are given full attention in the recent textbook "Regression Analysis in Medical Research 2nd Edition, Springer Heidelberg Germany, 2021, by the same authors."

This chapter particularly addresses, that, as an alternative, kernel ridge regressions may not only be less complex, and, therefore, more convenient, but also more sensitive. The same data example is used for comparison of different methodologies.

16.3 Data Example

The effect of the levels of satisfaction with the doctor on the levels of quality of life (qol) is assessed. In 450 patients with coronary artery disease the patient satisfaction with their doctor was assumed to be an important predictor of patient qol (quality of life). The datafile is entitled "qol", and is in SpringerLink supplementary files.

450 pts
Variable

1	2	3	4
qol	treatment	counseling	sat doctor
4	3	1	4
2	4	0	1
5	2	1	4
4	3	0	4
2	2	1	1
1	2	0	4
4	4	0	1
4	3	0	1
4	4	1	4
3	2	1	4

1. qol   = quality of life score (1 = very low, 5 = very high)
2. treatment = treatment modality (1 = cardiac fitness, 2 = physiotherapy, 3 = wellness, 4 = hydrotherapy, 5 = nothing)
3. counseling = counseling given (0 = no, 1 = yes)
4. sat doctor = satisfaction with doctor (1 = very low, 5 = very high).

The above table gives the first 10 patients of a 450 patients study of the effects of doctors' satisfaction level and qol. The entire data file is in SpringerLink supplementary files, and is entitled "qol". Start by downloading and opening the data file in your computer with SPSS statistical software 2022 version 28 0.1.0 installed.

16.4 Traditional Linear Regression

Command:

Menu....Analyze....Regression.....Linear....Dependent: qol score....Independent (s); counseling, satisfaction with doctor, treatment ....click OK.

Model Summary

Model	R	R Square	Adjusted R Square	Std. Error of the Estimate
1	,325[a]	,105	,099	1,383

a. Predictors: (Constant), counseling, sat with doctor, treatment

ANOVA[b]

Model		Sum of Squares	df	Mean Square	F	Sig.
1	Regression	100,559	3	33,520	17,526	,000[a]
	Residual	853,005	446	1,913		
	Total	953,564	449			

a. Predictors: (Constant), counseling, sat with doctor, treatment
b. Dependent Variable: qol score

Coefficients[a]

Model		Unstandardized Coefficients		Standardized Coefficients	t	Sig.
		B	Std. Error	Beta		
1	(Constant)	2,071	,231		8,980	,000
	sat with doctor	,244	,053	,207	4,615	,000
	treatment	,003	,056	,002	,056	,956
	counseling	,729	,131	,250	5,583	,000

a. Dependent Variable: qol score

The output shows an R Square of 0,105, which indicates 10,5% certainty to predict the outcome qol score. This is a poor result. R Square values under 0,25 are assumed to have very poor predictive potential.

16.5 Multinomial Regression

A more appropriate analysis for categorical outcome data may be multinomial regression. We will subsequently assess the effect of the predictors on the categorial outcome qol (quality of life scores) with multinomial regression.

Command:

Analyze....Regression....Multinomial Logistic Regression....Dependent: enter qol....
Factor(s): enter treatment, counseling, sat (satisfaction) with doctor....click OK.

The underneath tables are in the output. Although usual R Square values are
missing in the multinomial logistic model, Cox, Snell and Nagelkerke have devel-
oped alternative methods for estimating the proportion of certainty of the output
determined by the predictors and based on likelihood statistics. They called them
pseudo R Squares and have magnitudes similar to those of the real R Squares like
kernel ridge R Squares. The usually applied Cox and Snell R Square is larger than
the R Square computed by the traditional linear regression (0,191 vs 0,105).

Pseudo R-Square

Cox and Snell	,191
Nagelkerke	,199
McFadden	,067

Likelihood Ratio Tests

Effect	Model Fitting Criteria -2 Log Likelihood of Reduced Model	Likelihood Ratio Tests		
		Chi-Square	df	Sig.
Intercept	483,058[a]	,000	0	.
satdoctor	525,814	42,756	16	,000
treatment	496,384	13,326	12	,346
counseling	523,215	40,157	4	,000

The chi-square statistic is the difference in -2 log-likelihoods
between the final model and a reduced model. The reduced model
is formed by omitting an effect from the final model. The null
hypothesis is that all parameters of that effect are 0.

a. This reduced model is equivalent to the final model
   because omitting the effect does not increase the degrees
   of freedom.

Parameter Estimates

qol score[a]		B	Std. Error	Wald	df	Sig.	Exp(B)	95% Confidence Interval for Exp (B) Lower Bound	Upper Bound
very low	Intercept	-1,795	,488	13,528	1	,000			
	[treatment=1]	-,337	,420	,644	1	,422	,714	,314	1,626
	[treatment=2]	,573	,442	1,678	1	,195	1,773	,745	4,216
	[treatment=3]	,265	,428	,385	1	,535	1,304	,564	3,015
	[treatment=4]	0[b]	.	.	0	.			
	[counseling=0]	1,457	,328	19,682	1	,000	4,292	2,255	8,170
	[counseling=1]	0[b]	.	.	0	.			
	[satdoctor=1]	2,035	,695	8,579	1	,003	7,653	1,961	29,871
	[satdoctor=2]	1,344	,494	7,413	1	,006	3,834	1,457	10,089
	[satdoctor=3]	,440	,468	,887	1	,346	1,553	,621	3,885
	[satdoctor=4]	,078	,465	,028	1	,867	1,081	,435	2,687
	[satdoctor=5]	0[b]	.	.	0	.			
low	Intercept	-2,067	,555	13,879	1	,000			
	[treatment=1]	-,123	,423	,084	1	,771	,884	,386	2,025
	[treatment=2]	,583	,449	1,684	1	,194	1,791	,743	4,320
	[treatment=3]	-,037	,462	,006	1	,936	,964	,389	2,385
	[treatment=4]	0[b]	.	.	0	.			
	[counseling=0]	,846	,323	6,858	1	,009	2,331	1,237	4,392
	[counseling=1]	0[b]	.	.	0	.			
	[satdoctor=1]	2,735	,738	13,738	1	,000	15,405	3,628	65,418
	[satdoctor=2]	1,614	,581	7,709	1	,005	5,023	1,607	15,698
	[satdoctor=3]	1,285	,538	5,704	1	,017	3,614	1,259	10,375
	[satdoctor=4]	,711	,546	1,697	1	,193	2,036	,699	5,933
	[satdoctor=5]	0[b]	.	.	0	.			
medium	Intercept	-1,724	,595	8,392	1	,004			
	[treatment=1]	-,714	,423	2,858	1	,091	,490	,214	1,121
	[treatment=2]	,094	,438	,046	1	,830	1,099	,465	2,594
	[treatment=3]	-,420	,459	,838	1	,360	,657	,267	1,615
	[treatment=4]	0[b]	.	.	0	.			
	[counseling=0]	,029	,323	,008	1	,929	1,029	,546	1,940
	[counseling=1]	0[b]	.	.	0	.			
	[satdoctor=1]	3,102	,790	15,425	1	,000	22,244	4,730	104,594
	[satdoctor=2]	2,423	,632	14,714	1	,000	11,275	3,270	38,875
	[satdoctor=3]	1,461	,621	5,534	1	,019	4,309	1,276	14,549
	[satdoctor=4]	1,098	,619	3,149	1	,076	2,997	,892	10,073
	[satdoctor=5]	0[b]	.	.	0	.			
high	Intercept	-,333	,391	,724	1	,395			
	[treatment=1]	-,593	,371	2,562	1	,109	,552	,267	1,142
	[treatment=2]	-,150	,408	,135	1	,713	,860	,386	1,916
	[treatment=3]	,126	,376	,113	1	,737	1,135	,543	2,371
	[treatment=4]	0[b]	.	.	0	.			
	[counseling=0]	-,279	,284	,965	1	,326	,756	,433	1,320
	[counseling=1]	0[b]	.	.	0	.			
	[satdoctor=1]	1,650	,666	6,146	1	,013	5,208	1,413	19,196
	[satdoctor=2]	1,263	,451	7,840	1	,005	3,534	1,460	8,554
	[satdoctor=3]	,393	,429	,842	1	,359	1,482	,640	3,432
	[satdoctor=4]	,461	,399	1,337	1	,248	1,586	,726	3,466
	[satdoctor=5]	0[b]	.	.	0	.			

The above table are also in the output and they show that the multinomial regression computes for each outcome category, as compared to a defined reference category, the effect of each of the predictor variables separately without taking interactions into account. The ordinary regression, in contrast, computes the effect of the separate predictors on a single outcome as a variable consisting of categories with a stepping pattern. For analysis the statistical model Ordinal Regression in the SPSS module Regression is required. The above example is used once more.

16.6 Ordinal Regression

Command:

Analyze....Regression....Ordinal Regression....Dependent: enter qol....Factor(s): enter "treatment", "counseling", "sat with doctor"....click Options....Link: click Complementary Log-log....click Continue....click OK.

Pseudo R-Square

Cox and Snell	,117
Nagelkerke	,122
McFadden	,039

Link function: Logit.

Model Fitting Information

Model	-2 Log Likelihood	Chi-Square	df	Sig.
Intercept Only	578,352			
Final	537,075	41,277	8	,000

Link function: Complementary Log-log.

Parameter Estimates

		Estimate	Std. Error	Wald	df	Sig.	95% Confidence Interval	
							Lower Bound	Upper Bound
Threshold	[qol = 1]	-2,207	,216	103,925	1	,000	-2,631	-1,783
	[qol = 2]	-1,473	,203	52,727	1	,000	-1,871	-1,075
	[qol = 3]	-,959	,197	23,724	1	,000	-1,345	-,573
	[qol = 4]	-,249	,191	1,712	1	,191	-,623	,124
Location	[treatment=1]	,130	,151	,740	1	,390	-,167	,427
	[treatment=2]	-,173	,153	1,274	1	,259	-,473	,127
	[treatment=3]	-,026	,155	,029	1	,864	-,330	,277
	[treatment=4]	0[a]	.	.	0	.	.	.
	[counseling=0]	-,289	,112	6,707	1	,010	-,508	-,070
	[counseling=1]	0[a]	.	.	0	.	.	.
	[satdoctor=1]	-,947	,222	18,214	1	,000	-1,382	-,512
	[satdoctor=2]	-,702	,193	13,174	1	,000	-1,081	-,323
	[satdoctor=3]	-,474	,195	5,935	1	,015	-,855	-,093
	[satdoctor=4]	-,264	,195	1,831	1	,176	-,646	,118
	[satdoctor=5]	0[a]	.	.	0	.	.	.

Link function: Complementary Log-log.

a. This parameter is set to zero because it is redundant.

The above tables are in the output of the ordinal regression. Again a pseudo R Square value is produced. Although the model fitting information table tells, that the ordinal model provides an excellent overall fit for the data, the overall Cox and Snell pseudo R Square of the ordinal model is only 0,117, even smaller than that of the multinomial regression (0,191).

16.7   Kernel Ridge Regression

Finally kernel ridge regressions were performed.

Command:

Menu....Analyze....Regression....Kernel Ridge Regression....Dependent: qol....Independent(s): satdoctor, treatment, counseling....click Linear....click OK.

The table below is in the output. It shows a poor result of the of the linear kernel ridge density model. The R Square was even negative, −0,056. For the explanation of negative R Squares see Chap. 5 "Some Terminology".

Model Summary[a,b]

Kernel	Alpha	R Square
Linear	1,000	-,056

a. Dependent Variable: qol

b. Model: satdoctor,
   treatment, counseling

More kernel regressions were performed with similar commands.

Model Summary[a,b]

Kernel	Alpha	Gamma	R Square
Chi2	1,000	1,000	,241

a. Dependent Variable: qol

b. Model: satdoctor, treatment,
   counseling

Model Summary[a,b]

Kernel	Alpha	R Square
Cosine	1,000	,030

a. Dependent Variable: qol

b. Model: satdoctor,
   treatment, counseling

Model Summary[a,b]

Kernel	Alpha	Gamma	R Square
Laplacian	1,000	,333	,262

a. Dependent Variable: qol

b. Model: satdoctor, treatment, counseling

Model Summary[a,b]

Kernel	Alpha	Gamma	Coef0	Degree	R Square
Polynomial	1,000	,333	1,000	3,000	,205

a. Dependent Variable: qol

b. Model: satdoctor, treatment, counseling

Model Summary[a,b]

Kernel	Alpha	Gamma	R Square
RBF	1,000	,333	,237

a. Dependent Variable: qol

b. Model: satdoctor, treatment, counseling

Model Summary[a,b]

Kernel	Alpha	Gamma	Coef0	R Square
Sigmoid	1,000	,333	1,000	-,199

a. Dependent Variable: qol

b. Model: satdoctor, treatment, counseling

The results of the above kernel ridge density models show, that, particularly.

the Chi2 kernel density model (R Square 0,241),
the Laplacian kernel density model (R Square 0,262),
the Polynomial kernel density model (R Square 0,205),
the RBF kernel density model (R Square 0,237)

performed (much) better than did the linear kernel density model. A kernel density R Square over 0,250 like the one with the Laplacian model is assumed to be a reasonable result in terms of predictive performance. Note that the above analyses use four different kernel density models. Graphical presentations of them are in the Chap. 5 entitled "Some Terminology".

16.8   Conclusions

16.8.1   Summaries of Traditional Regressions

The R Square value of the traditional linear regression was 0,105, which indicates 10,5% certainty to predict the outcome qol score. This is a poor result. R Square values under 0,25 are assumed to have very poor predictive potential. The usually applied Cox and Snell R Square of the multinomial regression was larger than the R Square computed by the traditional linear regression (0,191 vs 0,105). The overall Cox and Snell pseudo R Square of the ordinal model is only 0,117, even smaller than that of the multinomial regression (0,191).

16.8.2   Summaries of Kernel Density Regressions

Kernel ridge density models show, that, particularly

the Chi2 kernel density model (R Square 0,241),
the Laplacian kernel density model (R Square 0,262),
the Polynomial kernel density model (R Square 0,205),
the RBF kernel density model (R Square 0,237)

performed (much) better than did the above traditional regression methodologies. A kernel density R Square over 0,250 like the one of the Laplacian model is assumed to be a reasonable result in terms of predictive performance.

16.9   References

All of the chapters of the current edition start with a brief review of the traditional analytic method of the different regression methods prior to the review of the relevant kernel ridge regression method. For the purpose, generally, data examples are used from the recent edition "Regression Analyses in Clinical Research for Starters and 2nd Levelers 2nd Edition, Springer Heidelberg Germany 2021", by the same authors. For a better understanding of differences between traditional and alternative regressions, readers may benefit from the study of this edition first.

To readers requesting still more background, theoretical and mathematical information of computations given, several textbooks complementary to the current production and written by the same authors are available: Statistics applied to clinical studies 5th edition, 2012, Machine learning in medicine a complete overview, 2015, SPSS for starters and 2nd levelers 2nd edition, 2015, Clinical data analysis on a pocket calculator 2nd edition, 2016, Understanding clinical data analysis from published research, 2016, all of them edited by Springer Heidelberg Germany.

Chapter 17
Effect of Department and Patient-Age on Risk of Falling out of Bed, 55 Patients, Traditional Regression vs Kernel Ridge Regression

Abstract As a 55 patient study example, three hospital departments (no surgery, little surgery, lot of surgery), and three patient age classes (young, middle, old) were the predictors of the risk of three classes of falling out of bed (fall out of bed:

"no",
"yes but no injury",
"yes and injury").

The multinomial model shows, that the department is a very significant predictor of fallingoutofbed, and also ageclass is a significant factor with p-values <0.000 compatible with the very high pseudo R squares of 0,718 (Cox and Snell) and 0,594(McFadden). With kernel ridge regressions the best fit kernel density model was the Polynomial model with R Square 0,771. More well fitting kernel density R Square values were the

Additive_chi2 kernel ridge density model with R square 0,748,
Chi2 kernel ridge density model with R square 0,756,
RBF (radial basis function) kernel ridge density model with
R square 0, 756.

Keywords Multinomial regression · Pseudo R squares · Polynomial kernel density · Additive chi2 kernel density · Chi2 kernel density · RBF kernel density

17.1 Summary

Outcome categories can be assessed with multinomial logistic regression. With categories both in the outcome and as predictors, the latter models assume, that for each predictor category or combination of categories x_1, x_2,... slightly different

---

Supplementary Information The online version contains supplementary material available at [https://doi.org/10.1007/978-3-031-10717-7_17].

a-values are computed with a better fit for the outcome category y than a single a-value

$$y = a + b_1x_1 + b_2x_2 + \ldots.$$

We should add that, instead of the above linear equation, even better results may be obtained with log-transformed outcome variables (log = natural logarithm).

$$\log y = a + b_1x_1 + b_2x_2 + \ldots.$$

As a 55 patient study example, three hospital departments (no surgery, little surgery, lot of surgery), and three patient age classes (young, middle, old) were the predictors of the risk class of falling out of bed (fall out of bed:

"no",
"yes but no injury",
"yes and injury").

Kernel ridge regression can also handle binary or multinomial outcome data. And despite the dicho- or multitomous outcomes, no pseudo R Squares are needed, but, rather, real kernel ridge R Square values can be obtained.

17.1.1   Summary of Traditional Regression with Multinomial Logistic Regression

The multinomial model shows, that the department is a very significant predictor of fallingoutofbed, and also ageclass is a significant factor with p-values <0.000 compatible with the very high pseudo R squares of 0,718 and 0,594.

17.1.2   Summary of Kernel Ridge Regression

The best fit kernel density model was here the Polynomial with R Square 0,771. a linear model in scatterplot of "falloutofbed-data" by "predicted data" was observed in the kernel ridge Polynomial kernel density model.

More well fitting kernel density R Square values were the

Additive_chi2 kernel ridge density model with R square 0,748,
Chi2 kernel ridge density model with R square 0,756,
RBF (radial basis function) kernel ridge density model with
R square 0, 756.

17.2   Introduction

Outcome categories can be assessed with multinomial regression. As an example, in a study, three hospital departments (no surgery, little surgery, lot of surgery), and three patient age classes (young, middle, old) were the predictors of the risk class of falling out of bed (fall out of bed no, yes but no injury, yes and injury). Are the predictor categories significant determinants of the risk of falling out of bed with or without injury. Kernel ridge regression can also handle binary or multinomial outcome data. And despite the dicho- or multitomous outcomes, no pseudo R squares are needed, but, rather, real kernel ridge R Square values can be obtained.

17.3   Data Example

As a 55 patient study example, three hospital departments (no surgery, little surgery, lot of surgery), and three patient age classes (young, middle, old) were the predictors of the risk class of falling out of bed (fall out of bed:

"no",
"yes but no injury",
"yes and injury").

Only the first 10 patients of the 55 patient file is shown above. The entire data file is in SpringerLink supplementary data, and is entitled "multinomial". It was previously used by the authors in SPSS for starters and 2nd levelers, Chap. 45, Springer Heidelberg Germany, 2016. SPSS version 20 and up can be used for analysis.

outcome fall out of bed	predictor department	predictor ageclass	patient_id
1	0	1,00	1,00
1	0	1,00	2,00
1	0	2,00	3,00
1	0	1,00	4,00
1	0	1,00	5,00
1	0	,00	6,00
1	1	2,00	7,00
1	0	2,00	8,00
1	1	2,00	9,00
1	0	,00	10,00

department =	department class (0 = no surgery, 1 = little surgery, 2 = lot of surgery)
falloutofbed =	risk of falling out of bed (0 = fall out of bed no, 1 = yes but no injury, 2 = yes and injury)
ageclass =	patient age classes (young, middle, old)
patient_id =	patient identification

Only the first 10 patients of the 55 patient file is shown above. The entire data file is in SpringerLink supplementary data, and is entitled "multinomial". It was previously used by the authors in SPSS for starters and 2nd levelers, Chap. 45, Springer Heidelberg Germany, 2016. SPSS version 20 and up can be used for analysis. For kernel ridge regression in SPSS statistical software the 2022 version 28 0.1.0 is required.

17.4   Traditional Regression with Multinomial Logistic Regression

We will first perform a traditional multinomial logistic regression of the data from the data example. Start by opening the datafile downloaded from SpringerLink supplementary files in your computer mounted with the above SPSS statistical software program version 28 0.1.0 or many earlier versions.

Command:

Analyze....Regression....Multinomial Regression....Dependent: enter falloutofbed.... Factor(s): enter department, ageclass....click OK.

The output sheets show no R Squares. R Squares are never produced with logistic regressions. Instead likelihood ratio statistics with similar predictive meaning are applied for the purpose. They are called pseudo R Square values, and they are pretty good here with a pseudo R-Square of 0,718 for the Cox and Snell version and 0,594 for the McFadden version.

Pseudo R-Square

Cox and Snell	,718
McFadden	,594

Likelihood Ratio Tests

Effect	Model Fitting Criteria	Likelihood Ratio Tests		
	-2 Log Likeliho od of Reduce d Model	Chi-Square	df	Sig.
Intercept	19,317[a]	,000	0	
department	47,878	28,561	4	,000
agecat	41,237	21,920	4	,000

The chi-square statistic is the difference in -2 log-likelihoods between the final model and a reduced model. The reduced model is formed by omitting an effect from the final model. The null hypothesis is that all parameters of that effect are 0.

a. This reduced model is equivalent to the final model because omitting the effect does not increase the degrees of freedom.

Parameter Estimates

fall with/out injury[a]		B	Std. Error	Wald	df	Sig.	Exp(B)	95% Confidence Interval for Exp (B)	
								Lower Bound	Upper Bound
no	Intercept	20,112	1,399	206,577	1	,000			
	[department=0]	-1,926	1,858	1,075	1	,300	,146	,004	5,562
	[department=1]	-19,833	1,092	329,841	1	,000	2,436E-9	2,865E-10	2,071E-8
	[department=2]	0[b]	.	.	0	.	.	.	.
	[agecat=,00]	-39,606	,000	.	1	.	6,297E-18	6,297E-18	6,297E-18
	[agecat=1,00]	17,101	,981	303,884	1	,000	26722844,16	3907110,550	1,828E8
	[agecat=2,00]	0[b]	.	.	0	.	.	.	.
yes/ but no injury	Intercept	17,804	1,061	281,502	1	,000			
	[department=0]	2,009	1,314	2,337	1	,126	7,453	,568	97,869
	[department=1]	-15,766	,000	.	1	.	1,422E-7	1,422E-7	1,422E-7
	[department=2]	0[b]	.	.	0	.	.	.	.
	[agecat=,00]	-21,423	,000	.	1	.	4,969E-10	4,969E-10	4,969E-10
	[agecat=1,00]	17,402	,000	.	1	.	36118618,14	36118618,14	36118618,14
	[agecat=2,00]	0[b]	.	.	0	.	.	.	.

a. The reference category is: yes/ and injury.
b. This parameter is set to zero because it is redundant.

The multinomial model shows, that the department is a very significant predictor of fallingoutofbed, and also ageclass is a significant factor with p-values <0,000 compatible with the very high pseudo R squares of 0,718 and 0,594.

17.5   Kernel Ridge Regression

Can we obtain even better accuracy, reliability and precision of testing using kernel ridge regressions? Kernel ridge regression can also handle binary or multinomial outcome data. And despite the dicho- or multitomous outcomes, no pseudo R Squares are needed, but, rather, real kernel ridge R Square values can be obtained. We will start with the linear kernel ridge density model. Remember that the statistical software in your computer must be SPSS statistical software 2022 version 28.0.1.0. Start by mounting this program in your computer and opening the datafile onto this software.

Command then:

Menu....Analyze....Regression....Kernel   Ridge   Regression....   Dependent: falloutofbed.... Independent(s): department, agecat....click Linear....click OK.

In the output the underneath table is given. Like many times before we observe that the computed R Square value is very negative. The explanation is given in the Chap. 5 entitled "Some Terminology". Obviously, the linear density model fits the dataset very poorly.

Model Summary[a,b]

Kernel	Alpha	R Square
Linear	1,000	-,871

a. Dependent Variable: falloutofbed

b. Model: department, agecat

We will assess more kernel ridge density model by similar commands given.

Command:

Menu....Analyze....Regression....Kernel   Ridge   Regression....   Dependent: falloutofbed....   Independent(s):   department,   agecat....click   Linear....click Additive_chi2....click OK.

Model Summary[a,b]

Kernel	Alpha	R Square
Additive_chi2	1,000	,748

a. Dependent Variable:
   falloutofbed

b. Model: department, agecat

Model Summary[a,b]

Kernel	Alpha	Gamma	R Square
Chi2	1,000	1,000	,756

a. Dependent Variable: falloutofbed

b. Model: department, agecat

Model Summary[a,b]

Kernel	Alpha	R Square
Cosine	1,000	-,795

a. Dependent Variable:
   falloutofbed

b. Model: department,
   agecat

Model Summary[a,b]

Kernel	Alpha	Gamma	Coef0	Degree	R Square
Polynomial	1,000	,500	1,000	3,000	,771

a. Dependent Variable: falloutofbed

b. Model: department, agecat

Model Summary[a,b]

Kernel	Alpha	Gamma	R Square
RBF	1,000	,500	,756

a. Dependent Variable: falloutofbed

b. Model: department, agecat

Model Summary[a,b]

Kernel	Alpha	Gamma	Coef0	R Square
Sigmoid	1,000	,500	1,000	,341

a. Dependent Variable: falloutofbed

b. Model: department, agecat

The best fit kernel density model was here the Polynomial with R Square 0,771. We will draw graphs and compute predicted values by the Polynomial kernel density model.

Command:

Analyze....Regression....Kernel Ridge Regression....Dependent: falloutofbed.... Independent(s): department, agecat....click Polynomial....click Options....mark Observed vs. Predicted....mark Residuals vs. Predicted....mark Predicted values.... click OK.

The underneath scatterplot is in the output. It may be kind of hard, but a very good linear predictive model can be recognized from the patterns in the plot.

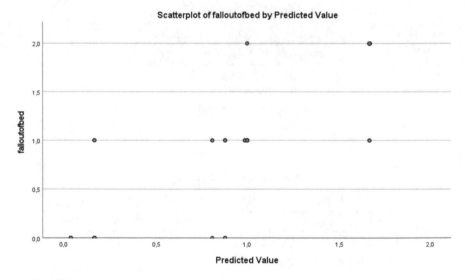

The linear model in the above scatterplot gives the "falloutofbed-data" by "predicted data" as computed by the kernel ridge Polynomial kernel density model.

More well fitting kernel density R Square values were

the Additive_chi2 kernel ridge density model with R Square 0,748
the Chi2 kernel ridge density model with R Square 0,756
the RBF kernel ridge density model with R Square 0, 756

These kernel ridge R square values were better than the pseudo R Square values from the traditional multinomial logistic regression analysis of the same dataset, namely 0,718 (and 0,594).

17.6 Conclusions

Outcome categories can be assessed with multinomial logistic regression. With categories both in the outcome and as predictors, the latter models assume, that for each predictor category or combination of categories x_1, x_2,... slightly different a-values are computed with a better fit for the outcome category y than a single a-value.

$$y = a + b_1 x_1 + b_2 x_2 + \ldots.$$

We should add that, instead of the above linear equation, even better results may be obtained with log-transformed outcome variables (log = natural logarithm).

$$\log y = a + b_1 x_1 + b_2 x_2 + \ldots.$$

As an 55 patient study example, three hospital departments (no surgery, little surgery, lot of surgery), and three patient age classes (young, middle, old) were the predictors of the risk class of falling out of bed (fall out of bed no, yes but no injury, yes and injury).

Kernel ridge regression can also handle binary or multinomial outcome data. And despite the dicho- or multitomous outcomes, no pscudo R Squares are needed, but, rather, real kernel ridge R Square values can be obtained.

17.6.1 Traditional Regression with Multinomial Logistic Regression

The multinomial model shows, that the department is a very significant predictor of fallingoutofbed, and also ageclass is a significant factor with p-values <0.000 compatible with the very high pseudo R squares of 0,718 and 0,594.

17.6.2   Kernel Ridge Regressions

The best fit kernel density model was here the Polynomial with R Square 0,771. a linear model in scatterplot of "falloutofbed-data" by "predicted data" was observed in the kernel ridge Polynomial kernel density model.

More well fitting kernel density R Square values were the

Additive_chi2 kernel ridge density model with R Square 0,748,
Chi2 kernel ridge density model with R Square 0,756,
RBF kernel ridge density model with R Square 0, 756.

Note, that the above seven analyses use seven different kernel density models. Graphical presentations of them are in the Chap. 5 entitled "Some Terminology".

17.7   References

All of the chapters of the current edition start with a brief review of the traditional analytic method of the different regression methods prior to the review of the relevant kernel ridge regression method. For the purpose, generally, data examples are used from the recent edition "Regression Analyses in Clinical Research for Starters and 2nd Levelers 2nd Edition, Springer Heidelberg Germany 2021", by the same authors. For a better understanding of differences between traditional and kernel regressions, readers may benefit from the study of this edition first.

To readers requesting still more background, theoretical and mathematical information of computations given, several textbooks complementary to the current production and written by the same authors are available: Statistics applied to clinical studies 5th edition, 2012, Machine learning in medicine a complete overview, 2015, SPSS for starters and 2nd levelers 2nd edition, 2015, Clinical data analysis on a pocket calculator 2nd edition, 2016, Understanding clinical data analysis from published research, 2016, all of them edited by Springer Heidelberg Germany.

Chapter 18
Effect of Diet, Gender, Sport, and Medical Treatment on LDL Cholesterol Reduction, 953 Patients, Traditional Regression vs Kernel Ridge Regression

Abstract In 953 patients various predictors of LDL (low density lipoprotein) - cholesterol reduction included weight reduction, gender, sport, treatment level, diet.

The traditional ANOVA(analysis of variance) R-square was 0,972, corresponding with a very good predictive model. In the coefficients table of the traditional ordinary least regression analysis, only weight reduction and sport activities were very significant predictors. Two methods are relevant for further assessment. In the regression-tree only weight reduction significantly contributed to the model ($p = 0,094$), with the overall mean (and standard deviation) as dependent variable ldl-cholesterol reduction. With kernel ridge regressions the best fit predictive model was given by the Laplacian kernel density model with an R Square value of 0,981. And so, decision tree analysis contributed little significance to the overall analysis.

Keywords Analysis of variance (anova) · Decision trees · Laplacian kernel density modeling

18.1 Summary

With decision trees, data samples of patients with and without the presence of a disease are assessed for subgroup properties. Usually binary variables are used for assessment, but binary cut-off values of continuous variables can also be used. Linear regression can be applied for finding the optimal cut-offs of subgroups, i.e., the cut-offs with the linear regression equation, that produces the largest test statistic. It is a lot of work, and it is, sometimes, called exhaustive searching, but for a computer it is not hard to do. We should add, that the computer is even capable of finding best cut-offs with partitioning into more than two subgroups, if required. But we will stick to two subgroups, in order to avoid too much complexity in the models.

Supplementary Information The online version contains supplementary material available at [https://doi.org/10.1007/978-3-031-10717-7_18].

Instead of or in addition to regression-tree-analysis quantile regressions may be helpful to provide additional significances of the predictor variables unobserved in the ordinary least squares regression and/or in the entire tree regression model. A 953 patient data file is used of various predictors of LDL (low density lipoprotein) - cholesterol reduction including weight reduction, gender, sport, treatment level, diet. The weight reduction and sport were very significant independent predictors of LDL cholesterol reduction.

18.1.1   Summaries of the Traditional Multiple Variables Linear Regressions

The traditional R-square is 0,972, a very good predictive model corresponding with an ANOVA (analysis of variance) with an F test of 662,847 and a p-value <0,000. In the coefficients table of the traditional ordinary least regression analysis, particularly weight reduction and sport activities were very significant predictors, three more independent variable were insignificant. Probably, interactions between the five independent variable were responsible. Two methods are relevant for further assessment. In the regression tree only weight reduction significantly contributed to the model (p = 0,094), with the overall mean and standard deviation dependent variable ldl-cholesterol reduction as dependent variable.

18.1.2   Summaries of Kernel Ridge Regressions

The best fit predictive model was given by the Laplacian kernel density model with an R Square value of 0,981. A very poor predictive model was the sigmoid kernel density model with a very large negative R Square value (see the Chap. 5 "Some Terminology" for explanation of negative R Square values).terol in the parent (root) node. And so, decision tree analysis contributed little significance to the overall analysis. The scatterplot of the Laplacian kernel ridge density model shows a very close linear pattern. A nice linear scatterplot gives the predicted values vs the predicted residuals. When returning to the data view screen, a 7th data column giving the predicted ldl_reductions by the Laplacian kernel ridge density model, is observed.

18.2   Introduction

With decision trees, data samples of patients with and without the presence of a disease are assessed for subgroup properties. Usually binary variables are used for assessment, but binary cut-off values of continuous variables can also be used. Linear regression can be applied for finding the optimal cut-offs of subgroups, i.e., the cut-offs with the linear regression equation, that produces the largest test statistic. It is a lot of work, and it is, sometimes, called exhaustive searching, but for a computer it is not hard to do. We should add, that the computer is even capable of finding best cut-offs with partitioning into more than two subgroups, if required. But we will stick to two subgroups, in order to avoid too much complexity in the models. Instead of or in addition to regression-tree-analysis kernel ridge regressions may be helpful to provide additional significances of the predictor variables unobserved in the ordinary least squares regression and in the tree regression model.

18.3 Data Example

A 953 patient data file is used of various predictors of LDL (low-density-lipoprotein)-cholesterol reduction including weight reduction, gender, sport, treatment level, diet. The SPSS data file is in SpringerLink supplementary files and is entitled "exhaustive testing". It is previously used by the authors in Machine learning in medicine a complete overview, Chap. 53, Springer Heidelberg Germany, 2015. The file is opened in your computer with SPSS statistical software installed. The first 13 patients are given underneath.

Variables (Var)					
1	2	3	4	5	6
3,41	0	1	3,00	3	0
1,86	-1	1	2,00	3	1
,85	-2	1	1,00	4	1
1,63	-1	1	2,00	3	1
6,84	4	0	4,00	2	0
1,00	-2	0	1,00	3	0
1,14	-2	1	1,00	3	1
2,97	0	1	3,00	4	0
1,05	-2	1	1,00	4	1
,63	-2	0	1,00	3	0
1,18	-2	0	1,00	2	0
,96	-2	1	1,00	2	0
8,28	5	0	4,00	2	1

Var 1 ldl_reduction
Var 2 weight_reduction (kg)
Var 3 gender (0-1)
Var 4 sport (1-4)
Var 5 treatment_level (1-4)
Var 6 diet (0-1)

18.4   Traditional Linear Regression

In your computer installed with SPSS statistical software 2022 version 28 0.1.0 download from SpringerLink supplementary files the data file entitled "exhausted testing". Start by opening the data file in the software program.

Command:

Analyze....Regression analysis....Linear....Dependent: ldl reduction....Independent (s): weight reduction, gender, sport, treatment level, diet....click OK.

In the output are the underneath tables. The traditional R-square is 0,972, a very good predictive model corresponding with an ANOVA (analysis of variance) with an F test of 662,847 and a p-value <0,000.

The underneath tables are coming up.

Model Summary

Model	R	R Square	Adjusted R Square	Std. Error of the Estimate
1	,986[a]	,972	,972	,36045

a. Predictors: (Constant), diet, treatment level, gender, weight reduction, sport

ANOVA[b]

Model		Sum of Squares	df	Mean Square	F	Sig.
1	Regression	4328,402	5	865,680	6662,847	,000[a]
	Residual	123,040	947	,130		
	Total	4451,443	952			

a. Predictors: (Constant), diet, treatment level, gender, weight reduction, sport
b. Dependent Variable: ldl reduction

Coefficients[a]

Model		Unstandardized Coefficients		Standardized Coefficients	t	Sig.
		B	Std. Error	Beta		
1	(Constant)	2,786	,065		43,088	,000
	weight reduction	,985	,012	,944	81,937	,000
	gender	-,015	,023	-,003	-,633	,527
	sport	,095	,023	,047	4,060	,000
	treatment level	,009	,010	,005	,954	,340
	diet	-,017	,023	-,004	-,716	,474

a. Dependent Variable: ldl reduction

In the coefficients table of the traditional ordinary least regression analysis, particularly weight reduction and sport activities were very significant predictors, three more independent variable were insignificant. Probably, interactions between the five independent variable were responsible. Two methods are relevant for further assessment.

First, decision tree analysis, using classification trees can sometimes reveal statistically significant results, although the multiple testing issue is a serious problem with this methodology. Second, kernel ridge regression is a relevant methodology, because it can deal with multicollinearity in a special way. How does ridge regression deal with multicollinearity? Ridge Regression is a technique for analyzing multiple regression data that suffer from multicollinearity. **By adding a degree of bias to the regression estimates**, ridge regression reduces the standard errors. It is hoped, that the net effect will then be to give estimates that are more reliable and less overfitted.

18.4.1 Decision Tree Analysis (Exhaustive Testing)

We will assess the effect of weight reduction on ldl cholesterol reduction, and try and find the best cut-off for assessment. First, we will transform the continuous weight reduction variable into various binary variables with different levels of cut-offs and use linear regression with ldl reduction as outcome for identifying the cut-off with the largest test statistic assessed with analysis of variance.

Command:

in Data View click weight_reduction....click Transform....click Compute Variable.... in Numeric Expression: enter weight_reduction<= 0....in Target Variable enter weight_reduction2....click OK....in Variable View under Name enter weightreduction2.... click Analyze....Regression....Linear.... Independent Variables: enter weight_reduction....Dependent Variable: enter ldl_reduction....click OK.

The same commands have to be executed for different weight reduction cut-offs, for example:

weight_reduction<= 0
weight_reduction<= 1
weight_reduction<= 2
weight_reduction<= ....

In the output sheets of the regressions the underneath F-statistics are given.

with weight_reduction<= 0 the F-statistic equals   1992,721
with weight_reduction<= 1   2963,858
with weight_reduction<= 2   2267,989.

Obviously, weight reduction $< = 1$ produces the best statistic, and is, thus, the cut-off level, which provides the most significant difference between ldl reductions, that can be obtained from differences in weight reductions. And, so, it is the best level for making predictions about the effect of weight reduction on ldl reduction.

We will now perform an automated entire regression tree analysis from the ldl-cholesterol example.

Command:

Analyze. . . .Classify. . .Tree. . . . Dependent Variable: enter ldl_reduction. . . . Independent Variables: enter weight reduction, gender, sport, treatment level, diet . . . . Growing Methods: select CRT . . . .click Criteria: enter Parent Node 300, Child Node 100. . . .click Output: Tree mark Tree in table format. . . .click OK.

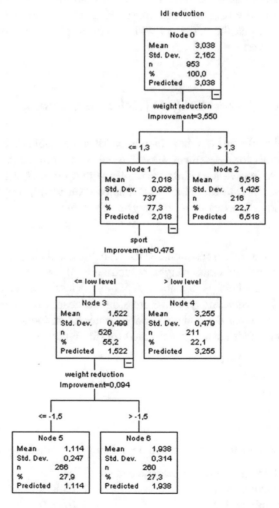

The output sheets show the above regression tree. Only weight reduction significantly contributed to the model (p = 0,094), with the overall mean and standard deviation dependent variable ldl-cholesterol in the parent (root) node. And so, decision tree analysis contributed little significance to the overall analysis.

18.5   Kernel Ridge Regression

Kernel ridge regression of the effect on ldl reduction of various predictors is assessed with the help of various kernel density models. In the 2022 version of SPSS statistical software 28.0.1.0 is kernel ridge models available. Start by opening the above datafile again.

Command:

Analyze....Regression....Kernel Ridge Regression....Dependent: ldl_reduction.... Independent(s): weight_reduction, gender, sport, treatment_level, diet....click Linear....click OK.

In the output sheets is the R Square value of the Linear kernel density model given. A very strong predictive magnitude of 91,8% certainty about the outcome ldl reduction by all of the predictors adjusted for multicollinearity.

Model Summary[a,b]

Kernel	Alpha	R Square
Linear	1,000	,918

a. Dependent Variable:
   ldl_reduction

b. Model: weight_reduction,
   gender, sport,
   treatment_level, diet

We will, subsequently, assess more kernel density models. Giving similar commands for the purpose.

Model Summary[a,b]

Kernel	Alpha	R Square
Cosine	1,000	,773

a. Dependent Variable:
   ldl_reduction

b. Model: weight_reduction,
   gender, sport,
   treatment_level, diet

Model Summary[a,b]

Kernel	Alpha	Gamma	R Square
Laplacian	1,000	,200	,981

a. Dependent Variable: ldl_reduction

b. Model: weight_reduction, gender, sport, treatment_level, diet

Model Summary[a,b]

Kernel	Alpha	Gamma	Coef0	Degree	R Square
Polynomial	1,000	,200	1,000	3,000	,978

a. Dependent Variable: ldl_reduction

b. Model: weight_reduction, gender, sport, treatment_level, diet

Model Summary[a,b]

Kernel	Alpha	Gamma	R Square
RBF	1,000	,200	,974

a. Dependent Variable: ldl_reduction

b. Model: weight_reduction, gender, sport, treatment_level, diet

Model Summary[a,b]

Kernel	Alpha	Gamma	Coef0	R Square
Sigmoid	1,000	,200	1,000	-113907,726

a. Dependent Variable: ldl_reduction

b. Model: weight_reduction, gender, sport, treatment_level, diet

The best fit predictive model was given by the Laplacian kernel density model with an R Square value of 0,981. A very poor predictive model was the sigmoid kernel density model with a very large negative R Square value (see Chap. 5 "Some Terminology" for explanation of negative R Square values).

The best fit kernel density model was thus provided by the Laplacian kernel density version with an R Square value of 0,981 We will try and draw graphs of the predicted ldl reductions of this kernel density model.

Command:

Analyze....Regression....Kernel Ridge Regression....Dependent: ldl_reduction.... Independent(s): weight_reduction, gender, sport, treatment_level, diet....click

Laplacian....click   Options....mark   Observed   vs.   Predicted....mark
Residuals vs. Predicted....mark Predicted values....click OK.

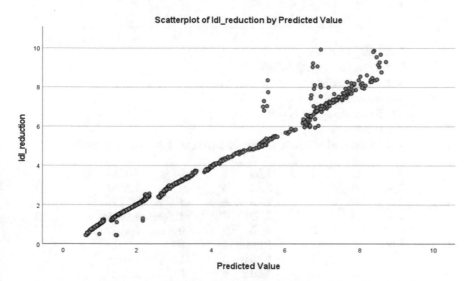

The above scatterplot of the Laplacian kernel ridge density model shows a very
close linear pattern.

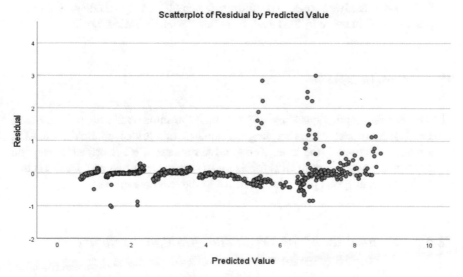

The above nice linear scatterplot gives the predicted values vs the predicted
residuals.

When returning to the data view screen of your computer you will observe a 7th
data column giving the predicted ldl_reductions by the Laplacian kernel ridge
density model. The first 10 patient data of the 953 patient datafile are in the
underneath table.

Variables (Var)						
1	2	3	4	5	6	7

Var 1 ldl_reduction
Var 2 weight_reduction (kg)
Var 3 gender (0-1)
Var 4 sport (1-4)
Var 5 treatment_level (1-4)
Var 6 diet (0-1)
Var 7 predicted ldl reduction by the Laplacian kernel density model

3,41	0	1	3,00	3	0	3,3653
1,86	-1	1	2,00	3	1	1,8567
,85	-2	1	1,00	4	1	,8717
1,63	-1	1	2,00	3	1	1,6295
6,84	4	0	4,00	2	0	6,9707
1,00	-2	0	1,00	3	0	,9891
1,14	-2	1	1,00	3	1	1,0710
2,97	0	1	3,00	4	0	2,9383
1,05	-2	1	1,00	4	1	1,0274
,63	-2	0	1,00	3	0	,7145

The above analyses used six different kernel density models. Graphical presnetations of them are in the Chap. 5 entitled "Some Terminology".

18.6   Conclusions

The kernel ridge regressions showed very much the same as did the ordinary least squares (OLS) model with very large R squares of various predictors on LDL cholesterol reduction all the time. Kernel ridge regression is a worthwhile analytic model for additional use with OLS (ordinary least square regressions). Decision tree regression modeling was pretty inconclusive in the example given.

18.6.1   Summaries of the Traditional Multiple Variables Linear Regressions

The traditional R-square is 0,972, a very good predictive model corresponding with an ANOVA (analysis of variance) with an F test of 662,847 and a p-value <0,000. In the coefficients table of the traditional ordinary least regression analysis, particularly weight reduction and sport activities were very significant predictors, three more independent variable were insignificant. Probably, interactions between the five independent variable were responsible. Two methods are relevant for further assessment. In the regression tree only weight reduction significantly contributed to the model ($p = 0,094$), with the overall mean and standard deviation dependent variable ldl-cholesterol reduction as dependent variable.

18.6.2 Summaries of Kernel Ridge Regressions

The best fit predictive model was given by the Laplacian kernel density model with an R Square value of 0,981. A very poor predictive model was the sigmoid kernel density model with a very large negative R Square value (see Chap. 5 "Some Terminologies" for explanation of negative R Square values).terol in the parent (root) node. And so, decision tree analysis contributed little significance to the overall analysis.

The scatterplot of the Laplacian kernel ridge density model shows a very close linear pattern. A nice linear scatterplot gives the predicted values vs the predicted residuals. When returning to the data view screen, a 7th data column giving the predicted ldl_reductions by the Laplacian kernel ridge density model is observed.

18.7 References

All of the chapters of the current edition start with a brief review of the traditional analytic method of the different regression methods prior to the review of the relevant kernel ridge regression method. For the purpose, generally, data examples are used from the recent edition "Regression Analyses in Clinical Research for Starters and 2nd Levelers 2nd Edition, Springer Heidelberg Germany 2021", by the same authors. For a better understanding of differences between traditional and alternative regressions, readers may benefit from the study of this edition first.

To readers requesting still more background, theoretical and mathematical information of computations given, several textbooks complementary to the current production and written by the same authors are available: Statistics applied to clinical studies 5th edition, 2012, Machine learning in medicine a complete overview, 2015, SPSS for starters and 2nd levelers 2nd edition, 2015, Clinical data analysis on a pocket calculator 2nd edition, 2016, Understanding clinical data analysis from published research, 2016, all of them edited by Springer Heidelberg Germany.

Chapter 19
Effect of Gender, Age, Weight, and Height on Measured Body Surface, 90 Patients, Traditional Regression vs Kernel Ridge Regression

Abstract The effect on measured body surface of predictors including gender, age, weight, and height was assessed in 90 patients. The traditional linear regression with four predictors produced an R Square of no less than 0,996. Thus, the predictors together explained the outcome by 99,6%. A very good result. The insignificant gender effect and the weakly significant age effect might have been caused by multicollinearities or interactions. As kernel ridge regression adjusts for multicollinearity, we will now assess kernel ridge regressions of the same data.

With kernel ridge regressions the polynomial kernel density model provided an R Square 0,997 (99,7% predictive certainty). More very strong predicting kernel density models were observed, namely the Linear, the Additive_chi2, and the Chi2 kernel density models.

Keywords Multicollinearities · Interactions · Polynomial kernel densities · Linear kernel densities · Additive_chi2 kernel densities · Chi2 kernel densities

19.1 Summary

In the current chapter the effect on measured body surface of predictors including gender, age, weight, and height were assessed in 90 patients.

19.1.1 Summaries of the Traditional Multiple Variables Linear Regression

The traditional linear regression with four predictors produced an R Square of no less than 0,996. Thus, the predictors together explained the outcome by 99,6%. A very

Supplementary Information The online version contains supplementary material available at [https://doi.org/10.1007/978-3-031-10717-7_19].

good result. The insignificant gender effect and the weakly significant age effect might have been caused by multicollinearities or interactions. As kernel ridge regression adjusts for multicollinearity, we will now assess kernel ridge regressions of the same data.

19.1.2   Summaries of the Kernel Ridge Regression

A negative R square like observed above is sometimes observed with kernel ridge density modeling. See also the Chap. 5 entitled "Some Terminology". In the underneath scatterplot the predicted values using the best fit kernel density model, the Polynomial kernel density model (R Square 0,997). More very strong predicting kernel density models were observed, the Linear, the Additive_chi2, and the Chi2 kernel density models. A very good linear data pattern is observed above with a Polynomial kernel density model. It is also seen that residuals have a constant pattern as it should, and do not increase with increasing predicted values. When returning to the dataview screen it is observed, that SPSS has provide predicted values for all of the patients.

19.2   Introduction

With traditional regression methods, the outcome values are assumed to be normally distributed around the regression line/curve. With non-normal outcome data, that remain non-normal in spite of transformations (Likert scales is a notorious example), data distributions may be skewed, and nonparametric regression analysis may provide better data fit than traditional parametric models do. Methods including nonparametric regression are pretty new, and not yet widely applied. They include: kriging, otherwise called Gaussian process regression, decision trees, and bagged (bootstrap aggregated) regression trees, kernel regressions, and median regression, otherwise called robust regression. In this chapter the effect of various predictors on measured body surfaces was studied in 90 persons. As methods of analysis traditional linear regression will be compared with kernel ridge regression analysis.

19.3   Data Example

The body surfaces of the 90 persons were calculated using direct photometric measurements. The data file, also used in the Chap. 1, is entitled "measured bodysurface", and is in SpringerLink supplementary files. The first 20 patients are in the underneath table.

Variables
Var (= variable)

1	2	3	4	5
1,00	13,00	30,50	138,50	10072,90
0,00	5,00	15,00	101,00	6189,00
0,00	0,00	2,50	51,50	1906,20
1,00	11,00	30,00	141,00	10290,60
1,00	15,00	40,50	154,00	13221,60
0,00	11,00	27,00	136,00	9654,50
0,00	5,00	15,00	106,00	6768,20
1,00	5,00	15,00	103,00	6194,10
1,00	3,00	13,50	96,00	5830,20
0,00	13,00	36,00	150,00	11759,00
0,00	3,00	12,00	92,00	5299,40
1,00	0,00	2,50	51,00	2094,50
0,00	7,00	19,00	121,00	7490,80
1,00	13,00	28,00	130,50	9521,70
1,00	0,00	3,00	54,00	2446,20
0,00	0,00	3,00	51,00	1632,50
0,00	7,00	21,00	123,00	7958,80
1,00	11,00	31,00	139,00	10580,80
1,00	7,00	24,50	122,50	8756,10
1,00	11,00	26,00	133,00	9573,00

Var 1 = gender
Var 2 = age (years)
Var 3 = weight (kg)
Var 4 = height (m)
Var 5 = body surface measured (cm^2)

First, traditional multiple variables linear regression will be used, second, kernel ridge regressions will be applied.

19.4   Traditional Linear Regression

start by downloading and opening the data file entitled "bodysurface measured" from the SpringerLink site into your computer installed with SPSS statistical software 2022 version 28 0.1.0.

Command:

Analyze....Menu....Regression....Linear Regression....Dependent:enter bodysurface measured....Independent (s): enter gender, age, weight, height....click OK.

In the output sheets are the underneath tables.

Model Summary

Model	R	R Square	Adjusted R Square	Std. Error of the Estimate
1	,998[a]	,996	,996	209,16040

a. Predictors: (Constant), height, gender, weight, age

ANOVA[b]

Model		Sum of Squares	df	Mean Square	F	Sig.
1	Regression	1,019E9	4	2,547E8	5822,647	,000[a]
	Residual	3718586,183	85	43748,073		
	Total	1,023E9	89			

a. Predictors: (Constant), height, gender, weight, age
b. Dependent Variable: bodysurface measured

Coefficients[a]

Model		Unstandardized Coefficients		Standardized Coefficients	t	Sig.
		B	Std. Error	Beta		
1	(Constant)	-902,173	188,613		-4,783	,000
	gender	43,694	44,645	,006	,979	,331
	age	-42,802	20,490	-,063	-2,089	,040
	weight	171,143	8,130	,623	21,052	,000
	height	46,931	2,893	,446	16,224	,000

a. Dependent Variable: bodysurface measured

The traditional linear regression with four predictors produced an R Square of no less than 0,996. Thus, the predictors together explained the outcome by 99,6%. A very good result. The insignificant gender effect and the weakly significant age effect might have been caused by multicollinearities or interactions. As kernel ridge regression adjusts for multicollinearity, we will now assess kernel ridge regressions of the same data.

19.5   Kernel Ridge Regression

The advantage of the kernel ridge method is that, unlike with traditional multiple variables linear regression, multicollinearity is adjusted.

Command:

Analyze....Regression....Kernel Ridge Regression....Dependent (s): bodysurface measured....Independent(s): gender, age, weight, height....click Linear....click OK.

Model Summary[a,b]

Kernel	Alpha	R Square
Linear	1,000	,995

a. Dependent Variable: VAR00005

b. Model: VAR00001, VAR00002, VAR00003, VAR00004

The above table is in the output and shows that the linear kernel ridge regression model has a very strong datafit for making predictions about the outcome variable, bodysurface measured, equally strong as that of the traditional linear regression (R Square 0,996).

More kernel density models are available and will be applies next. Similar commands are given for the purpose.

Model Summary[a,b]

Kernel	Alpha	R Square
Additive_chi2	1,000	,997

a. Dependent Variable: VAR00005

b. Model: VAR00001, VAR00002, VAR00003, VAR00004

Model Summary[a,b]

Kernel	Alpha	Gamma	R Square
Chi2	1,000	1,000	,871

a. Dependent Variable: VAR00005

b. Model: VAR00001, VAR00002, VAR00003, VAR00004

Model Summary[a,b]

Kernel	Alpha	R Square
Cosine	1,000	,498

a. Dependent Variable: VAR00005

b. Model: VAR00001, VAR00002, VAR00003, VAR00004

Model Summary[a,b]

Kernel	Alpha	Gamma	R Square
Laplacian	1,000	,250	,554

a. Dependent Variable: VAR00005

b. Model: VAR00001, VAR00002, VAR00003, VAR00004

Model Summary[a,b]

Kernel	Alpha	Gamma	Coef0	Degree	R Square
Polynomial	1,000	,250	1,000	3,000	,997

a. Dependent Variable: VAR00005

b. Model: VAR00001, VAR00002, VAR00003, VAR00004

Model Summary[a,b]

Kernel	Alpha	Gamma	R Square
RBF	1,000	,250	,009

a. Dependent Variable: VAR00005

b. Model: VAR00001, VAR00002, VAR00003, VAR00004

Model Summary[a,b]

Kernel	Alpha	Gamma	Coef0	R Square
Sigmoid	1,000	,250	1,000	-,001

a. Dependent Variable: VAR00005

b. Model: VAR00001, VAR00002, VAR00003, VAR00004

A negative R square like observed above is sometimes observed with kernel ridge density modeling. See also the Chap. 5 entitled "Some Terminology".

In the underneath scatterplot is the predicted values using the best fit kernel density model, the Polynomial kernel density model (R Square 0,997). More very strong predicting kernel density models were observed, the Linear, the Additive_chi2, and the Chi2 kernel density models.

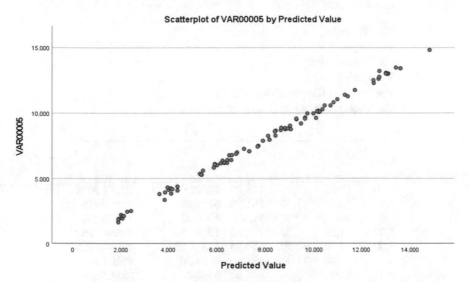

A very good linear data pattern is observed above with a Polynomial kernel density model. Underneath it is also seen, that residuals have a constant pattern as it should, and do not increase with increasing predicted values.

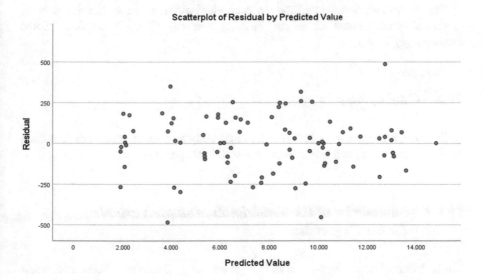

When returning to the dataview screen it is observed, that SPSS has provided predicted values for all of the patients. Only the first 10 patients are shown in the underneath table.

Var = variable

Var 1gender
Var 2 age
Var 3 weight (kg)
Var 4 height (m)
Var 5 body surface measured (cm^2)
Var 6 body surface predicted (cm^2)

Var 1	2	3	4	5	6
1,00	13,00	30,50	138,50	10072,90	10208,6094
,00	5,00	15,00	101,00	6189,00	6306,1016
,00	,00	2,50	51,50	1906,20	1927,5410
1,00	11,00	30,00	141,00	10290,60	10354,7969
1,00	15,00	40,50	154,00	13221,60	12734,5625
,00	11,00	27,00	136,00	9654,50	9620,6563
,00	5,00	15,00	106,00	6768,20	6514,9688
1,00	5,00	15,00	103,00	6194,10	6428,9727
1,00	3,00	13,50	96,00	5830,20	5882,4766
,00	13,00	36,00	150,00	11759,00	11717,2969

The above kernel ridge regressions used eight different kernel density models. Graphical presentations of all of them are in the Chap. 5 entitled "Some Terminology".

19.6   Conclusions

In the current chapter the effect on measured body surface of predictors including gender, age, weight, and height were assessed in 90 patients.

19.6.1   Summaries of the Traditional Multiple Variables Linear Regression

The traditional linear regression with four predictors produced an R Square of no less than 0,996. Thus, the predictors together explained the outcome by 99,6%. A very good result. The insignificant gender effect and the weakly significant age effect might have been caused by multicollinearities or interactions. As kernel ridge regression adjusts for multicollinearity, we will now assess kernel ridge regressions of the same data.

19.6.2   Summaries of the Kernel Ridge Regression

A negative R square like observed above is sometimes observed with kernel ridge density modeling. See also the Chap. 5 "Some Terminology".

In the underneath scatterplot the predicted values using the best fit kernel density model, the Polynomial kernel density model (R Square 0,997). More very strong predicting kernel density models were observed, the Linear, the Additive_chi2, and the Chi2 kernel density models.

A very good linear data pattern is observed above with a Polynomial kernel density model. It is also seen that residuals have a constant pattern as it should, and do not increase with increasing predicted values. When returning to the dataview screen it is observed that SPSS has provide predicted values for all of the patients.

19.7   References

All of the chapters of the current edition start with a brief review of the traditional analytic method of the different regression methods prior to the review of the relevant kernel regression method. For the purpose, generally, data examples are used from the recent edition "Regression Analyses in Clinical Research for Starters and 2nd Levelers 2nd Edition, Springer Heidelberg Germany 2021", by the same authors. For a better understanding of differences between traditional and other regressions, readers may benefit from the study of this edition first.

To readers requesting still more background, theoretical and mathematical information of computations given, several textbooks complementary to the current production and written by the same authors are available: Statistics applied to clinical studies 5th edition, 2012, Machine learning in medicine a complete overview, 2015, SPSS for starters and 2nd levelers 2nd edition, 2015, Clinical data analysis on a pocket calculator 2nd edition, 2016, Understanding clinical data analysis from published research, 2016, all of them edited by Springer Heidelberg Germany.

Chapter 20
Effect of Physicians' Characteristics on Their Inclination to Give Lifestyle Advise or Not, 139 Physicians, Traditional Regression vs Kernel Ridge Regression

Abstract In 139 physicians the effect on giving lifestyle advise or not of various predictors is assessed. Each physician produced two rows of data, one with and one without prior education as outcome. The predictor variables were physicians' identity, age, prior postgraduate education or not, and country-practice or not. Only the physician id and having had prior postgraduate education were significant predictors. The R square is not easily obtained from binary outcome data like logistic models, but, instead, SPSS provides pseudo R squares obtained from loglikelihoods. The Cox and Snell R square value was 0,579, which is a pretty good value for making reliable predictions. It means that 57,9% certainty for predicting the outcome should be provided by the statistical model given.

With kernel ridge regression the effect of physician education on lifestyle advise is visualized by a series of strong positive R square values in various kernel density models. The best fit model is the Chi2 kernel density model, with R square 0,865. The model provides 86.5% certainty by the predictors about the outcome. However, more kernel density models were strong predictors too.

Keywords Pseudo R squares · Logistic models · Loglikelihoods · Cox and Snell R square · Chi2 kernel density model

20.1 Summary

In 139 physicians the effect on giving lifestyle advise or not of various predictors is assessed.

Supplementary Information The online version contains supplementary material available at [https://doi.org/10.1007/978-3-031-10717-7_20].

20.1.1 Summaries of Traditional Multiple Variables Logistic Regressions

The variable "physician" is the physician identity number. Each physician produced two rows of data, one with and one without prior education as outcome. The predictor variables were physicians' identity, age, prior postgraduate education or not, and countrypractice or not. Only the physician id and having had prior postgraduate education were significant predictors.

The R square is not easily obtained from binary outcome data like logodds, but, instead, SPSS provides pseudo R squares obtained from loglikelihood statistics. The Cox and Snell R square value was 0,579, which is a pretty good value for making reliable predictions. It means that 57,9% certainty for predicting the outcome should be given by the statistical model given. The Nagelkerke pseudo R square is even better (0,785) although infrequently used.

20.1.2 Summaries of Kernel Ridge Regressions

The effect of physician education on lifestyle advise is visualized by a series of strong positive R square values in various kernel density models. The best fit model is the Chi2 kernel density model, with R square 0,865. The model provides 86.5% certainty by the predictors about the outcome. However, more kernel density models were strong predictors too. The Chi2 kernel density model R square of 0,865 (86,5% certainty) is much better than the pseudo R square value of the Cox and Snell approach of only 0,579, i.e., 57,9%.

More powerful kernel ridge regression R square predictors were:

Linear kernel ridge regression   R square 0,598
Additive_chi2 kernel ridge regression   R square 0,664
Laplacian kernel ridge regression   R square 0,750
Polynomial kernel ridge regression   R square 0,709
RBF (radial basis function) kernel ridge ridge regression   R square 0, 656.

All of them were better sensitive than the pseudo R square value of Cox and Snell of 0,579 as established in the traditional binary logistic regression analysis of the same data.

20.2 Introduction

The effect of various physicians' characteristics like age, prior postgraduate education, country practice, on giving lifestyle advise or not can be statistically tested using paired non-parametric mcNemar Chi-square tests. However, McNemar tests do not provide R Square values, and our objective was to assess with the help of R

Square values, whether Kernel ridge regressions are a more powerful alternative method of data analysis than traditional methods. As alternative to McNemar's chi-square tests, a binary logistic regression with physicians' id number as predictor and lifestyle as outcome is possible. In the current chapter we will use this method as traditional analysis, and test its predictive power against various kernel ridge kernel density methods.

20.3 Patient Example

In 139 physicians the effect on giving lifestyle advise or not of various predictors is assessed. The predictors were physicians' age, prior postgraduate education, countrypractice. Each physician has been tested twice in a crossover fashion, and in the summary statistics listed in two rows, one with prior postgraduate education, one without. The prior assumption was that no carryover effect between the first and second period of treatment was in the data. The first eight patients of the 139 patient dataset is given underneath. The remainder is in the datafile entitled "paireddata". Start by downloading the datafile from SpringerLink supplementary data into your computer mounted with SPSS statistical software 2022 version 28.0.1.0.

Variables				
1	2	3	4	5
physician id number	age	education yes/no	lifestyle advise	countrypractice yes/no
1	89,00	1	,00	1,00
1	89,00	2	,00	1,00
2	78,00	1	,00	1,00
2	78,00	2	,00	1,00
3	79,00	1	,00	1,00
3	79,00	2	,00	1,00
4	76,00	1	,00	1,00
4	76,00	2	,00	1,00
5	87,00	1	,00	1,00
5	87,00	2	,00	1,00
6	84,00	1	,00	1,00
6	84,00	2	,00	1,00
7	84,00	1	,00	1,00
7	84,00	2	,00	1,00
8	69,00	1	,00	1,00
8	69,00	2	,00	1,00

Variable 1 = physician id
Variable 2 = age (years)
Variable 3 = prior postgraduate education yes / no
Variable 4 = lifestyle advise given yes / no
Variable 5 = countrypractice yes / no

20.4   Traditional Regression (Binary Logistic Regression)

Although binary logistic regression is originally meant for testing unpaired data only, it might be a plausible alternative analysis method for paired data with a zero carryover effect.

Command:

Analyze....Regression....Binary Logistic Regression....dependent: Lifestyleadvise.... Covariates: physicianid, age, postgraduateeducation, countrypractice....click OK.
    The output sheets tables are underneath.

Model Summary

Step	-2 Log likelihood	Cox & Snell R Square	Nagelkerke R Square
1	130,985ᵃ	,579	,785

a. Estimation terminated at iteration number 7 because parameter estimates changed by less than ,001.

Classification Table[a]

			Predicted		
			lifestyleadvise		
Observed			,00	1,00	Percentage Correct
Step 1	lifestyleadvise	,00	157	13	92,4
		1,00	15	93	86,1
	Overall Percentage				89,9

a. The cut value is ,500

Variables in the Equation

		B	S.E.	Wald	df	Sig.	Exp(B)
Step 1ᵃ	physicianid	,098	,013	60,837	1	,000	1,103
	age	-,021	,017	1,634	1	,201	,979
	postgraduateeducation	1,579	,488	10,482	1	,001	4,848
	countrypractice	-,051	,240	,045	1	,831	,950
	Constant	-9,301	1,852	25,217	1	,000	,000

a. Variable(s) entered on step 1: physicianid, age, postgraduateeducation, countrypractice.

The variable "physician" is the physician identity number. Each physician produced two rows of data, one with and one without prior education as outcome. The predictor variables were physicians' identity, age, prior postgraduate education or not, and countrypractice or not. Only the physician id and having had prior postgraduate education were significant predictors.

The R Square is not easily obtained from binary outcome data like logodds, but, instead, SPSS provides pseudo R squares obtained from loglikelihood statistics. The Cox and Snell R square value was 0,579, which is a pretty good value for making reliable predictions. It means that 57,9% certainty for predicting the outcome should be given by the statistical model given. The Nagelkerke pseudo R square is even better (0,785) although infrequently used.

20.5 Kernel Ridge Regression

Kernel ridge regressions were performed of the effects of physicianid, age, prior postgraduate education or not, countrypractice or not on the outcome lifestyleadvise given or not.

Command:

Analyze....Regression....Kernel Ridge Regression....Dependent:lifestyleadvise.... Dependent(s): physicianid, age, postgraduateeducation, countrypractice....click Linear....click OK.

Model Summary[a,b]

Kernel	Alpha	R Square
Linear	1,000	,598

a. Dependent Variable: lifestyleadvise

b. Model: physicianid, age, postgraduateeducation, countrypractice

In the output sheets a kernel ridge linear regression is given. The R square value was 0,598 corresponding with a 59,8% predictive certainty of the predictors about the outcome, which is pretty good. It is also somewhat better than the pseudo R square of the traditional logistic regression Cox and Snell test of 0,579 (57,9% certainty). More kernel density models are available and were also assessed.

Command:

Analyze....Regression....Kernel Ridge Regression....Dependent: lifestyleadvise.... Dependent(s): physicianid, age, postgraduateeducation, countrypractice....click Linear....click Chi2....click OK.

Model Summary[a,b]

Kernel	Alpha	Gamma	R Square
Chi2	1,000	1,000	,865

a. Dependent Variable: lifestyleadvise

b. Model: physicianid, age, postgraduateeducation, countrypractice

The above table shows that the kernel density model Linear has been replaced with the kernel density model Chi2, and that a much better datafit has been obtained with now an R square value of 0,865 (86,5% certainty of prediction of outcome by the predictors).

More kernel density models were often pretty strong predictors. Six more of them were assessed, giving similar commands.

Model Summary[a,b]

Kernel	Alpha	R Square
Additive_chi2	1,000	,664

a. Dependent Variable: lifestyleadvise

b. Model: physicianid, age, postgraduateeducation, countrypractice

Model Summary[a,b]

Kernel	Alpha	R Square
Cosine	1,000	,463

a. Dependent Variable: lifestyleadvise

b. Model: physicianid, age, postgraduateeducation, countrypractice

Model Summary[a,b]

Kernel	Alpha	Gamma	R Square
Laplacian	1,000	,250	,750

a. Dependent Variable: lifestyleadvise

b. Model: physicianid, age,
   postgraduateeducation, countrypractice

Model Summary[a,b]

Kernel	Alpha	Gamma	Coef0	Degree	R Square
Polynomial	1,000	,250	1,000	3,000	,709

a. Dependent Variable: lifestyleadvise

b. Model: physicianid, age, postgraduateeducation, countrypractice

Model Summary[a,b]

Kernel	Alpha	Gamma	R Square
RBF	1,000	,250	,656

a. Dependent Variable: lifestyleadvise

b. Model: physicianid, age,
   postgraduateeducation,
   countrypractice

Model Summary[a,b]

Kernel	Alpha	Gamma	Coef0	R Square
Sigmoid	1,000	,250	1,000	,000

a. Dependent Variable: lifestyleadvise

b. Model: physicianid, age, postgraduateeducation,
   countrypractice

The effect of physician education on lifestyle advise is visualized by a series of strong positive R square values in various kernel density models. The best fit model is the Chi2 kernel density model, with R square 0,865. The model provides 86.5% certainty by the predictors about the outcome. However, more kernel density models were strong predictors too. The Chi2 kernel density model R square of 0,865 (86,5% certainty) is much better than the pseudo R square value of the Cox and Snell approach of only 0,579, i.e., 57,9%.

More powerful kernel ridge regression R square predictors were:

Linear kernel ridge regression	R square 0,598
Additive_chi2 kernel ridge regression	R square 0,664
Laplacian kernel ridge regression	R square 0,750

Polynomial kernel ridge regression   R square 0,709
RBF kernel ridge ridge regression   R square 0, 656.

All of them were better sensitive than the pseudo R square value of Cox and Snell of 0,579 as established in the traditional binary logistic regression analysis of the same data.

The above eight kernel ridge regressions used eight different kernel density models.

All of them are in the form of graphical presentations in the Chap. 5, entitled "Some Terminology".

20.6   Conclusion

20.6.1   Summaries of Traditional Multiple Variables Logistic Regressions

The variable "physician" is the physician identity number. Each physician produced two rows of data, one with and one without prior education as outcome. The predictor variables were physicians' identity, age, prior postgraduate education or not, and countrypractice or not. Only the physician id and having had prior postgraduate education were significant predictors.

The R square is not easily obtained from binary outcome data like logodds, but, instead, SPSS provides pseudo R squares obtained from loglikelihood statistics. The Cox and Snell R square value was 0,579, which is a pretty good value for making reliable predictions. It means that 57,9% certainty for predicting the outcome should be given by the statistical model given. The Nagelkerke pseudo R square is even better (0,785) although infrequently used.

20.6.2   Summaries of Kernel Ridge Regressions

The effect of physician education on lifestyle advise is visualized by a series of strong positive R Square values in various kernel density models. The best fit model is the Chi2 kernel density model, with R square 0,865. The model provides 86.5% certainty by the predictors about the outcome. However, more kernel density models were strong predictors too. The Chi2 kernel density model R square of 0,865 (86,5% certainty) is much better than the pseudo R square value of the Cox and Snell approach of only 0,579, i.e., 57,9%.

More powerful kernel ridge regression R square predictors were:

Linear kernel ridge regression   R square 0,598
Additive_chi2 kernel ridge regression   R square 0,664

Laplacian kernel ridge regression   R square 0,750
Polynomial kernel ridge regression   R square 0,709
RBF (radial basis function) kernel ridge ridge regression   R square 0, 656.

All of them were better sensitive than the pseudo R square value of Cox and Snell of 0,579 as established in the traditional binary logistic regression analysis of the same data.

20.7   References

All of the chapters of the current edition start with a brief review of the traditional analytic method of the different regression methods prior to the review of the relevant kernel ridge regression method. For the purpose, generally, data examples are used from the recent edition "Regression Analyses in Clinical Research for Starters and 2nd Levelers 2nd Edition, Springer Heidelberg Germany 2021", by the same authors. For a better understanding of differences between traditional and alternative regressions, readers may benefit from the study of this edition first.

To readers requesting still more background, theoretical and mathematical information of computations given, several textbooks complementary to the current production and written by the same authors are available: Statistics applied to clinical studies 5th edition, 2012, Machine learning in medicine a complete overview, 2015, SPSS for starters and 2nd levelers 2nd edition, 2015, Clinical data analysis on a pocket calculator 2nd edition, 2016, Understanding clinical data analysis from published research, 2016, all of them edited by Springer Heidelberg Germany.

Chapter 21
Effect of Treatment, Psychological and Social Scores on Numbers of Paroxysmal Atrial Fibrillations, 50 Patients, Traditional Regressions vs Kernel Ridge Regression

Abstract Fifty patients were studied for numbers of paroxysmal atrial fibrillations (PAFs), while on a parallel-group treatment with two different treatments. The scientific questions were: do psychological and social factor scores and different treatments affect the numbers of PAFs per person per period of time. A weighted least squares (WLS) regression provided a borderline significant effect of treatment modality on number of PAFs per person per period of time. The traditional linear regression provided an R square value of 0,603 meaning that the outcome can be predicted by 60%. A weighted least square method adjusted for the days of observation and should be a more precise analysis for our purpose. Unexpectedly, the adjustment caused the predictive property of the model to fall from R square = 0,603 to R square = 0,091. Certainty of prediction is only left by 9,1%, a very poor result in this analysis. Kernel ridge regressions provided sometimes much better fit data models (RBF = radial basis function):

Additive_chi2 kernel density model   R square 0,834
Chi2 kernel density model   R square 0,657
Laplacian kernel density model   R square 0,580
Polynomial kernel density model   R square 0,975
RBF kernel density model   R square 0,564.

Keywords Weighted least squares · R squares · Kernel density models

21.1 Summary

This chapter addressed the performance of kernel ridge regression versus traditional linear regression and a form of linear regression with time as weighting factor will be tested. As an example 50 patients were studied for numbers of paroxysmal atrial

Supplementary Information The online version contains supplementary material available at [https://doi.org/10.1007/978-3-031-10717-7_21].

fibrillations (PAFs), while on a parallel-group treatment with two different treatments. The scientific questions were: do psychological and social factor scores and different treatments affect the numbers of PAFs per person per period of time. A weighted least squares (WLS) regression provided a borderline significant effect of treatment modality on number of PAFs per person per period of time.

21.1.1   Summary of Traditional Regressions

Treatment modality is weakly significant, and psychological and social score are not. Furthermore, days of observation is very significant. However, it is not very precise to include this variable if your outcome is the numbers of events per person per time unit. Therefore, we will adjust the outcome variable for the differences in days of observation using weighted least square (WLS) regression.

The traditional linear regression provided an R square value of 0,603, pretty good, and meaning that the outcome can be predicted by the predictors by over 60%. The weighted least square method adjusts for the days of observation and is a more precise analysis for our purpose. Unfortunately, the adjustment caused the predictive property of the model to fall from R square = 0,603 to only R square = 0,091. Certainty of prediction is only left by 9,1%, a very poor result in this analysis.

21.1.2   Summary of Kernel Ridge Regressions

In order to assess, whether kernel ridge regression provided better fit data modeling different kernel density models were performed. Note, that SPSS version 28.0.1.0 is required, because earlier version of SPSS statistical software do not carry kernel ridge regressions. The best fit kernel density models were.

Additive_chi2 kernel density model   R square 0,834
Chi2 kernel density model   R square 0,657
Laplacian kernel density model   R square 0,580
Polynomial kernel density model   R square 0,975
RBF kernel density model   R square 0,564.

Of the above 5 kernel density models 3 were, thus, better sensitive predictors than were the traditional linear regression with an R Square value of 0,603 and weighted least square regression with an R square of 0,091.

21.2   Introduction

Linear regression measures events per population or person, but does not explicitly include time as a covariate, although, implicitly, it is often assumed, albeit not plainly expressed. In this chapter both traditional linear regression and a special form of linear regression with time as weighting factor will be tested against kernel ridge regressions.

21.3   Data Example

As an example 50 patients were studied for numbers of paroxysmal atrial fibrillations (PAFs), while on a parallel-group treatment with two different treatments. The Data file entitled "pafsperperson" is in SpringerLink supplementary files. The scientific questions were: do psychological and social factor scores and different treatments affect the numbers of PAFs per person per period of time. It is previously used by the authors in SPSS for starters and 2nd levelers, Chap. 21, Springer Heidelberg Germany, 2015. The first 9 patients are in the table underneath.

Variables				
1	2	3	4	5
PAFs	Treat	Psych	Soc	Days
4	1	56,99	42,45	73
4	1	37,09	46,82	73
2	0	32,28	43,57	76
3	0	29,06	43,57	74
3	0	6,75	27,25	73
13	0	61,65	48,41	62
11	0	56,99	40,74	66
7	1	10,39	15,36	72
10	1	50,53	52,12	63

Outcome variable
1. PAFs = counted numbers of paroxysmal atrial fibrillations per person.

Predictor variables
2. Treat = treatment modality
3. Psych = psychological score,
4. Soc = social score
5. Days = days of observation

SPSS statistical software will be used for analysis. Start with opening the data file in your computer that has SPSS installed. Then command.

21.4   Traditional Linear Regressions

Command:

Analyze. . . .Regression. . . .Linear. . . .Dependent Variable: episodes of paroxysmal atrial fibrillation. . . .Independent: treatment modality, psychological score, social score, days of observation. . . .click OK.

Model Summary

Model	R	R Square	Adjusted R Square	Std. Error of the Estimate
1	,777[a]	,603	,568	4,683

a. Predictors: (Constant), days observation, psychological sscore, treatment modality, social score

ANOVA[b]

Model		Sum of Squares	df	Mean Square	F	Sig.
1	Regression	1500,267	4	375,067	17,100	,000[a]
	Residual	987,013	45	21,934		
	Total	2487,280	49			

a. Predictors: (Constant), days observation, psychological sscore, treatment modality, social score
b. Dependent Variable: outcome rate

Coefficients[a]

Model		Unstandardized Coefficients		Standardized Coefficients	t	Sig.
		B	Std. Error	Beta		
1	(Constant)	49,059	5,447		9,006	,000
	treat	-2,914	1,385	-,204	-2,105	,041
	psych	,014	,052	,036	,273	,786
	soc	-,073	,058	-,169	-1,266	,212
	days	-,557	,074	-,715	-7,535	,000

a. Dependent Variable: paf

The above tables are in the output sheets and show, that treatment modality is weakly significant, and psychological and social score are not. Furthermore, days of observation is very significant. However, it is not very precise to include this variable if your outcome is the numbers of events per person per time unit. Therefore, we will perform a linear regression, and adjust the outcome variable for the differences in days of observation using weighted least square (WLS) regression.

Command: Analyze....Regression....Linear....Dependent: episodes of paroxysmal atrial fibrillation....Independent: treatment modality, psychological score, social score ....WLS Weight: days of observation....click OK.

Model Summary

Model	R	R Square	Adjusted R Square	Std. Error of the Estimate
1	,301[a]	,091	,031	53,190

a. Predictors: (Constant), social score, treatment modality, psychological sscore

ANOVA[b,c]

Model		Sum of Squares	df	Mean Square	F	Sig.
1	Regression	12962,913	3	4320,971	1,527	,220[a]
	Residual	130141,908	46	2829,172		
	Total	143104,820	49			

a. Predictors: (Constant), social score, treatment modality, psychological sscore
b. Dependent Variable: outcome rate
c. Weighted Least Squares Regression - Weighted by days observation

Coefficients[a,b]

Model		Unstandardized Coefficients		Standardized Coefficients	t	Sig.
		B	Std. Error	Beta		
1	(Constant)	10,033	2,862		3,506	,001
	treat	-3,502	1,867	-,269	-1,876	,067
	psych	,033	,069	,093	,472	,639
	soc	-,093	,078	-,237	-1,194	,238

a. Dependent Variable: paf

b. Weighted Least Squares Regression - Weighted by days

The above tables are in the output sheets of your computer. They shows the results. A largely similar pattern is observed, but treatment modality is no more statistically significant at $p = 0,050$.

21.5 Kernel Ridge Regression

The traditional linear regression provided an R square value of 0,603, pretty good, and meaning that the outcome can be predicted by the predictors by over 60%. The weighted least square method adjusts for the days of observation and may be a more precise analysis for our purpose. Unfortunately, the adjustment caused the predictive property of the model to fall from 0,603 to only 0,091. Certainty of prediction is only left by 9,1%, a very poor result in this analysis. In this section we will assess whether a better predictive model can be construed by kernel ridge regressions. We will start with the linear kernel ridge density model. Remember that the statistical software in your computer must be SPSS statistical software 2022 version 28.0.1.0. Start by mounting this program in your computer and download the data file into this software.

Command then:

Menu....Analyze....Regression....Kernel Ridge Regression.... Dependent: paf.... Independent(s): treat, psych, soc., days....click Linear....click OK.

In the output the underneath table is given. Like many times before we observe that the computed R square value is negative. The explanation is given in the Chap. 5 entitled "Some Terminology". Obviously the linear density model fitted the data very poorly.

Model Summary[a,b]

Kernel	Alpha	R Square
Linear	1,000	-,112

a. Dependent Variable: paf

b. Model: treat, psych, soc, days

We will assess more kernel ridge density model by similar commands given.

Model Summary[a,b]

Kernel	Alpha	R Square
Additive_chi2	1,000	,834

a. Dependent Variable: paf

b. Model: treat, psych, soc, days

Model Summary[a,b]

Kernel	Alpha	Gamma	R Square
Chi2	1,000	1,000	,657

a. Dependent Variable: paf

b. Model: treat, psych, soc, days

Model Summary[a,b]

Kernel	Alpha	R Square
Cosine	1,000	-,001

a. Dependent Variable: paf

b. Model: treat, psych, soc, days

Model Summary[a,b]

Kernel	Alpha	Gamma	R Square
Laplacian	1,000	,250	,580

a. Dependent Variable: paf

b. Model: treat, psych, soc, days

Model Summary[a,b]

Kernel	Alpha	Gamma	Coef0	Degree	R Square
Polynomial	1,000	,250	1,000	3,000	,975

a. Dependent Variable: paf

b. Model: treat, psych, soc, days

Model Summary[a,b]

Kernel	Alpha	Gamma	R Square
RBF	1,000	,250	,564

a. Dependent Variable: paf

b. Model: treat, psych, soc, days

Model Summary[a,b]

Kernel	Alpha	Gamma	Coef0	R Square
Sigmoid	1,000	,250	1,000	,000

a. Dependent Variable: paf

b. Model: treat, psych, soc, days

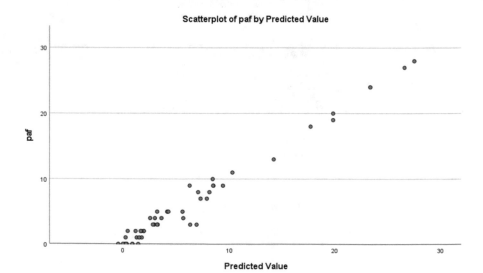

Scatterplot of paf by Predicted Value

With the kernel ridge polynomial density model applied, a very nice linear pattern of the scatterplot of pafs by predicted values is obtained.

The best fit kernel density models were.

Additive_chi2 kernel density model   R square 0,834
Chi2 kernel density model   R square 0,657
Laplacian kernel density model   R square 0,580
Polynomial kernel density model   R square 0,975
RBF kernel density model   R square 0,564.

Of the above five models three of them were better sensitive predictors than was the traditional linear regression with an R square value of 0,603.

The above analyses used eight different kernel density models. Graphical presentations of all of them are in the Chap. 5 entitled "Some Terminology".

21.6 Conclusion

The performance of kernel ridge regression was assessed against traditional linear regression and a form of linear regression with time as weighting factor entitled Weighted Least Squares (WLS) analysis. As example 50 patients were assessed for numbers of paroxysmal atrial fibrillations (PAFs), while on a parallel-group treatment with two different treatments. The scientific questions were: do psychological and social factor scores and different treatments affect the numbers of PAFs per person per period of time. A weighted least squares (WLS) regression provided a borderline significant effect of treatment modality on number of PAFs per person per period of time. In order to assess, whether kernel ridge regression provided equally adequate sensitivity of testing as did WLS, kernel ridge regressions were performed of different kernel ridge density models. Note, that SPSS version 28.0.1.0 is required here, because earlier versions of SPSS statistical software do not carry kernel ridge regression.

21.6.1 Summary of Traditional Regressions

Treatment modality is weakly significant, and psychological and social score are not. Furthermore, days of observation is very significant. However, it is not very precise to include this variable if your outcome is the numbers of events per person per time unit. Therefore, we will adjust the outcome variable for the differences in days of observation using weighted least square (WLS) regression.

The traditional linear regression provided an R square value of 0,603, pretty good, and meaning that the outcome can be predicted by the predictors by over 60%. The weighted least square method adjusts for the days of observation and is a more precise analysis for our purpose. Unfortunately, the adjustment caused the predictive property of the model to fall from R square $= 0{,}603$ to only R Square $= 0{,}091$. Certainty of prediction is only left by 9,1%, a very poor result in this analysis.

21.6.2   Summary of Kernel Ridge Regressions

In order to assess, whether kernel ridge regression provided better fit data modeling different kernel density models were performed. Note, that SPSS version 28.0.1.0 is required, because earlier version of SPSS statistical software do not carry kernel ridge regressions. The best fit kernel density models were.

Additive_chi2 kernel density model   R square 0,834
Chi2 kernel density model   R square 0,657
Laplacian kernel density model   R square 0,580
Polynomial kernel density model   R square 0,975
RBF kernel density model   R square 0,564.

Of the above 5 kernel density models 3 were, thus, better sensitive predictors than were the traditional linear regression with an R square value of 0,603 and weighted least square regression with an R square of 0,091.

21.7   References

All of the chapters of the current edition start with a brief review of the traditional analytic method of the different regression methods prior to the review of the relevant kernel ridge regression method. For the purpose, generally, data examples are used from the recent edition "Regression Analyses in Clinical Research for Starters and 2nd Levelers 2nd Edition, Springer Heidelberg Germany 2021", by the same authors. For a better understanding of differences between traditional and kernel regressions, readers may benefit from the study of this edition first.

To readers requesting still more background, theoretical and mathematical information of computations given, several textbooks complementary to the current production and written by the same authors are available: Statistics applied to clinical studies 5th edition, 2012, Machine learning in medicine a complete overview, 2015, SPSS for starters and 2nd levelers 2nd edition, 2015, Clinical data analysis on a pocket calculator 2nd edition, 2016, Understanding clinical data analysis from published research, 2016, all of them edited by Springer Heidelberg Germany.

Chapter 22
Effect of Various Predictors on Numbers of Convulsions in 3390 Patients, Traditional vs Kernel Ridge Regression

Abstract A data example of 3390 patients with epilepsy was studied for various predictors of convulsions. First a data model with three predictors was analyzed with traditional multiple variables linear regression, but the predictive potential was poor with an overall R square of only 0,003, and a p-value of p = 0,015. Second, a four predictor model was assessed. It performed slightly better with an R square of 0,007, and a p-value of <0,001. The kernel ridge regression of the four predictor model produced a much better R square value of 0,806 (80,6% predictive certainty) in the Laplacian kernel density model. A very good linear data fit of the predicted values versus the measured values (numbers of convulsions) was observed.

Keywords Traditional multiple variables linear regression · R-squares · Multiple variables kernel ridge regression · Laplacian kernel density model

22.1 Summary

A data example of 3390 patients with epilepsy was studied for various predictors of convulsions. The data file was partially obtained from the "SPSS Samples" site.

22.1.1 Summaries of Traditional Multiple Variables Linear Regressions

First a data model with three predictors was analyzed with traditional multiple variables linear regression, but the predictive potential was pretty poor with an overall R square of only 0,003, and a p-value of p = 0,015. Second, a four predictor

Supplementary Information The online version contains supplementary material available at [https://doi.org/10.1007/978-3-031-10717-7_22].

model was assessed. It performed slightly better with an R square of 0,007, and a p-value of <0,001.

22.1.2   Summaries of Kernel Ridge Linear Regressions

The kernel ridge regression of the four predictor model produced a much better R square value of 0,806 (80,6% predictive certainty) in the Laplacian kernel density model. A very good linear data fit of the predicted values versus the measured values (numbers of convulsions) was observed.

22.2   Introduction

A data example of 3390 patients with epilepsy was studied for various predictors of convulsions. The data file was partially obtained from the "SPSS Samples" site (138 kB). First a data model with three predictors was analyzed with traditional multiple variables linear regression, but the predictive potential was pretty poor with an overall R square of only 0,003, and a p-value of $p = 0,015$. Second, a four predictor model was assessed. It performed slightly better with an R square of 0,007, and a p-value of <0,001. The kernel ridge regression of the four predictor model produced a much better R square value of 0,806 (80,6% predictive certainty) in the Laplacian kernel density model.

22.3   Data Example

A data example of 3390 patients with epilepsy was studied for various predictors of convulsions. The data file was obtained from the SPSS Samples site (138 kB).

Variables

centersize	dob	treat	weeks	convulsions
Var 1	2	3	4	6
1	05/26/1990	0	0	2
1	05/26/1990	0	1	6
1	05/26/1990	0	2	4
1	05/26/1990	0	3	4
1	05/26/1990	0	4	6
1	05/26/1990	0	5	3
1	06/07/1977	1	0	4
1	06/07/1977	1	1	7
1	06/07/1977	1	2	5
1	06/07/1977	1	3	7
1	06/07/1977	1	4	6
1	06/07/1977	1	5	6

Variable 1 = Var 1
dob = day of birth
treat = treatment modality
weeks = weeks of observation
convulsions = number of convulsions

The data file entitled "anticonvulsants", stored for the purpose of the current edition at SpringerLink supplementary files, must be downloaded in your computer mounted with SPSS statistical software 2022 version 18.0.1.0 prior to self-assessment use. Start your analysis by opening the data file in your computer. We will first perform a traditional ordinary least square linear regression analysis, and then a kernel ridge regression with a Laplacian density model, because the linear kernel ridge density model provided very poor fit R squares of only 0,003 and 0,007.

22.4   Traditional Multiple Variables Regressions

Initially a three predictor multiple variables traditional linear regression model was analyzed.

Command:

click Analyze....Menu....Regression....Linear....Dependent: Convulsions....Independent(s): center size, dob (day of birth), weeks of observation....click OK.
   The underneath tables are in the output.

Model Summary

Model	R	R Square	Adjusted R Square	Std. Error of the Estimate
1	,056[a]	,003	,002	5,453

a. Predictors: (Constant), Week, Date of birth, Center size

ANOVA[b]

Model		Sum of Squares	df	Mean Square	F	Sig.
1	Regression	313,178	3	104,393	3,510	,015[a]
	Residual	100691,988	3386	29,738		
	Total	101005,167	3389			

a. Predictors: (Constant), Week, Date of birth, Center size
b. Dependent Variable: Number of convulsions

Coefficients[a]

Model		Unstandardized Coefficients		Standardized Coefficients	t	Sig.
		B	Std. Error	Beta		
1	(Constant)	12,131	2,766		4,386	,000
	Center size	-,033	,127	-,004	-,261	,794
	Date of birth	-5,483E-10	,000	-,042	-2,448	,014
	Week	-,115	,055	-,036	-2,106	,035

a. Dependent Variable: Number of convulsions

In the output sheets the overall R square = 0,003. The overall p-value = 0,015. This is a pretty poor result, and a four predictor alternative model was performed next.

Command:

Analyze....Menu....Regression....Linear....Dependent: Convulsions....Independent (s): center size, dob (date of birth), treatment yes/no, weeks of observation.... click OK.
The output sheets are underneath.

Model Summary

Model	R	R Square	Adjusted R Square	Std. Error of the Estimate
1	,083[a]	,007	,006	5,444

a. Predictors: (Constant), Week, Treatment received, Date of birth, Center size

ANOVA[a]

Model		Sum of Squares	df	Mean Square	F	Sig.
1	Regression	691,523	4	172,881	5,834	<,001[b]
	Residual	100313,644	3385	29,635		
	Total	101005,167	3389			

a. Dependent Variable: Number of convulsions

b. Predictors: (Constant), Week, Treatment received, Date of birth, Center size

Coefficients[a]

Model		Unstandardized Coefficients		Standardized Coefficients	t	Sig.
		B	Std. Error	Beta		
1	(Constant)	11,960	2,761		4,331	<,001
	Center size	-,045	,127	-,006	-,355	,723
	Date of birth	-5,601E-10	,000	-,043	-2,505	,012
	Treatment received	,669	,187	,061	3,573	<,001
	Week	-,115	,055	-,036	-2,110	,035

a. Dependent Variable: Number of convulsions

The output sheets show the above analysis tables of the traditional linear regression with four predictors. The R square $= 0.007$. The p-values for dob (date of birth), treatment received or not, and week of observation, were significant at $p < 0.05$.

22.5 Kernel Ridge Regression

Subsequently, a Laplacian kernel ridge regression will be performed.

Command:

Analyze....Regression....Kernel Ridge Regression....Dependent: Convulsions.... Independent(s): center size, date of birth, treatment received, weeks of observation....click Options: mark Observed vs Predicted....mark Predicted values....click Continue....click Laplacian....click OK.

The underneath tables and graph are in the output.

Model Summary[a,b]

Kernel	Alpha	Gamma	R Square
Laplacian	1,000	,250	,806

a. Dependent Variable: convulsions

b. Model: center_size, dob, treatment, week

A strong R square of 0,806 (80,6% predictive certainty) was obtained.

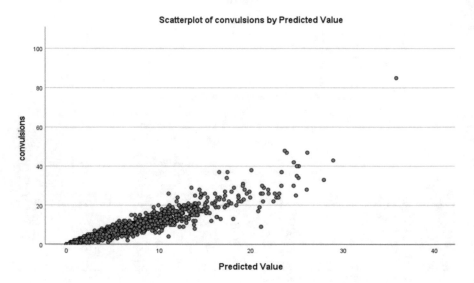

The above graph in the output showed a very good linear datafit of the predicted values versus the measured values (measured numbers of convulsions). The kernel ridge R square with a value of 0,806 was excellent as compared to that of the traditional regression, only 0,007.

After returning to the data sheets it can be observed that SPSS has added a novel column consistent of the program-predicted outcome values that are pretty close to the observed numbers of convulsions.

The underneath table shows these predicted values in the latter column. Only the data from first 10 subjects are in the table but all of the 3390 subject data are in the datasheets.

The first 10 subjects data file are in the underneath table.

center size	day of birth	treat	weeks	convulsions	predicted
1	0 05/26/1990	0	0	2	2,57
1	0 05/26/1990	0	1	6	3,58
1	0 05/26/1990	0	2	4	3,60
1	0 05/26/1990	0	3	4	3,65
1	0 05/26/1990	0	4	6	3,75
1	0 05/26/1990	0	5	3	2,94
1	1 06/07/1977	1	0	4	3,88
1	1 06/07/1977	1	1	7	4,73
1	1 06/07/1977	1	2	5	4,84
1	1 06/07/1977	1	3	7	5,28

We should add, that the above Laplacian kernel density model, as applied, is in graphical form in the Chap. 5, entitled "Some Terminology".

22.6 Conclusion

A data example of 3390 patients with epilepsy was studied for various predictors of convulsions. The data file was partially obtained from the "SPSS Samples" site (138 kB).

22.6.1 Summaries of Traditional Multiple Variables Linear Regressions

First a data model with three predictors was analyzed with traditional multiple variables linear regression, but the predictive potential was pretty poor with an overall R square of only 0,003, and a p-value of p = 0,015. Second, a four predictor model was assessed. It performed slightly better with an R square of 0,007, and a p-value of <0,001.

22.6.2 Summaries of Kernel Ridge Linear Regressions

The kernel ridge regression of the four predictor model produced a much better R square value of 0,806 (80,6% predictive certainty) in the Laplacian kernel density model. A very good linear data fit of the predicted values versus the measured values (numbers of convulsions) was observed.

22.7   References

All of the chapters of the current edition start with a brief review of the traditional analytic method of the different regression methods prior to the review of the relevant kernel ridge regression method. For the purpose, generally, data examples are used from the recent edition "Regression Analyses in Clinical Research for Starters and 2nd Levelers 2nd Edition, Springer Heidelberg Germany 2021", by the same authors. For a better understanding of differences between traditional and alternative regressions, readers may benefit from the study of this edition first.

To readers requesting still more background, theoretical and mathematical information of computations given, several textbooks complementary to the current production and written by the same authors are available: Statistics applied to clinical studies 5th edition, 2012, Machine learning in medicine a complete overview, 2015, SPSS for starters and 2nd levelers 2nd edition, 2015, Clinical data analysis on a pocket calculator 2nd edition, 2016, Understanding clinical data analysis from published research, 2016, all of them edited by Springer Heidelberg Germany.

Chapter 23
Effect of Foods Served on Breakfast Taken, 252 Persons, Traditional Linear and Multinomial Logistic Regression vs Kernel Ridge Regressions

Abstract A data example of 252 persons was studied for the effects of different foods on breakfast scenario taken. Traditional analyses consisted of analysis of variance and multinomial regression. They produced respectively an R Square value of 0,122 (12,2%) and a pseudo R Square of 0,400 (40%). With kernel ridge regressions the best data fits were provided by the Laplacian and the Polynomial kernel density models (respectively R square values 0,510 and 0,659, meaning 51,0% and 65,9% predictive certainties). The traditional multiple variables linear regression and multinomial model provided R square values of only 0,122 (12,2% predictive certainty), and 0,400 (40% certainty of prediction).

Keywords Analysis of variance · Multinomial regression · R Square · Pseudo R Square · Laplacian kernel densities · Polynomial kernel densities

23.1 Summary

A data example of 252 persons was studied for the effects of different foods on breakfast scenario taken.

23.1.1 Summaries of the Traditional Multiple Variables Linear Regression and Multinomial Logistic Regression

The R square value of 0,122 can be interpreted as a 12,2% certainty about the outcome given by all of the predictors simultaneously. This is less than 25%, and it means, that the model is a poor predictive model. The analysis of variance

Supplementary Information The online version contains supplementary material available at [https://doi.org/10.1007/978-3-031-10717-7_23].

(ANOVA) table produced a F (Fisher) statistic of 4236 compatible with a p-value of <0,001. As an alternative multinomial logistic regression may provide better fit results. With multinomial logistic regression, indeed various predictor variables were significant predictors of different outcome values, the menu scenarios. However, for comparisons with kernel ridge regression we are particularly interested in the overall R Square values, giving the percentage of certainty of the predictors on the outcome. However, traditional R Square values are not provided by the multinomial model. Fortunately, pseudo R Square based on log likelihood statistics are provided instead. Cox and Snell pseudo R Squares are commonly used. The value of 0,400 is a lot better fit than the R Square = 0,112 of the traditional linear regression of the data.

23.1.2   Summaries of the Kernel Ridge Regressions

The best data fits were provided by the Laplacian and the Polynomial kernel density models (respectively R square values 0,510 and 0,659, meaning 51,0% and 65,9% predictive certainties). The traditional multiple variables linear regression and multinomial model provided R square values of only 0,122 (12,2% predictive certainty), and 0,400 (40% predictive certainty.

Note, that the RBF (radial basis function) kernel density model produced a negative R square. Explanation is given in the Chap. 5 entitled "Some Terminology".

23.2   Introduction

The data file entitled "breakfast" is in SpringerLink supplementary files. This data file was partly taken from the "Samples site"of IBM SPSS Statistics 2022 Version 28.0.1.0. In 252 persons the effect of foods served on breakfast taken was studied. The outcome (Breakfast taken) was estimated as menu scenarios (1–6 different foods). The predictors were eight different foods. First, a traditional multiple variables linear regression was performed. Second, kernel ridge regressions were done using eight different kernel density models. The traditional multiple variables linear regression provided a p-value of <0,001, R square 0,122. The kernel ridge regression with linear kernel density model provided an R square of 0,066. Not a very powerful result, but in line with that of the traditional linear regression result. However, more kernel density models are available, and some of them provided much better R square values, sometimes even as large as 0,659. And so, kernel ridge regression is able to provide much better data fits than traditional linear regression did.

23.3 Data Example

The data file entitled "breakfast" is in SpringerLink supplementary files. This data file was partly taken from the "Samples site" of IBM SPSS Statistics 2022 Version 28.0.1.0. In 252 persons the effect of foods served on breakfast taken was studied. The outcome (Breakfast taken) was estimated as menu scenarios (1–6 different foods). The predictors were eight different foods.

First, a traditional multiple variables linear regression was performed. Second, kernel ridge regressions were done using eight different kernel density models. The first 12 patients of the data file is underneath. The remainder is of data is in SpringerLink supplementary files.

Variables 1 - 9

1	2	3	4	5	6	7	8	9
srcid	TP	BT	JD	TMd	TMn	CB	DP	CMB
1	13	12	3	11	15	2	1	14
1	15	11	3	8	12	7	1	13
1	15	10	14	8	11	1	6	13
1	6	14	3	10	15	4	1	13
1	15	9	14	8	10	11	1	5
1	9	11	4	10	12	5	2	13
1	9	14	6	11	15	7	2	10
1	15	10	6	8	11	3	1	14
1	15	12	4	11	13	7	9	14
1	15	13	7	12	14	5	2	3
1	9	2	15	10	7	11	13	3
1	11	1	15	8	14	10	9	6

Variable 1 = menu scenario (6 scrids, 1-6)
1 = overall preference
2 = breakfast with juice bacon eggs and beverage
3 = breakfast with juice cold cereal
4 = breakfast with juice pancakes sausage and beverage
5. = breakfast with beverage only
6 = snack with beverage only.

The predictor variables 2–9, TP, BT, JD, TMd, TMn, CB, DP, CMB, are successively listed underneath.

Toast pop-up
Buttered toast
Jelly donut
Toast and marmalade
Toast and margarine
Cinnamon bun
Danish pastry
Corn muffin and butter

23.4   Traditional Linear Regression and Multinomial Logistic Regression

First, a traditional linear regression with 8 predictors and menu scenarios as outcome was performed. Download the 252 person data file in your computer installed with SPSS statistical software 2022 version 28 0.1.0. Start by opening the data file.

Command:

Analyze....Menu....Regression....Linear Regression....Dependent: enter menu scenarios....Independent (s): enter all of the 8 predictors....click OK.
    In the output sheets are the underneath tables.

Model Summary

Model	R	R Square	Adjusted R Square	Std. Error of the Estimate
1	,350[a]	,122	,094	1,629

a. Predictors: (Constant), Corn muffin and butter, Toast and margarine, Toast pop-up, Toast and marmalade, Danish pastry, Jelly donut, Cinnamon bun, Buttered toast

ANOVA[a]

Model		Sum of Squares	df	Mean Square	F	Sig.
1	Regression	89,964	8	11,245	4,236	<,001[b]
	Residual	645,036	243	2,654		
	Total	735,000	251			

a. Dependent Variable: Menu scenarios

b. Predictors: (Constant), Corn muffin and butter, Toast and margarine, Toast pop-up, Toast and marmalade, Danish pastry, Jelly donut, Cinnamon bun, Buttered toast

Coefficients[a]

Model		Unstandardized Coefficients		Standardized Coefficients	t	Sig.
		B	Std. Error	Beta		
1	(Constant)	5,050	1,279		3,950	<,001
	Toast pop-up	-,045	,028	-,100	-1,592	,113
	Buttered toast	,086	,041	,226	2,065	,040
	Jelly donut	-,077	,033	-,200	-2,355	,019
	Toast and marmalade	,026	,030	,061	,871	,385
	Toast and margarine	-,056	,041	-,149	-1,370	,172
	Cinnamon bun	,009	,038	,020	,226	,821
	Danish pastry	-,037	,032	-,101	-1,169	,244
	Corn muffin and butter	-,053	,036	-,120	-1,471	,143

a. Dependent Variable: Menu scenarios

The R square value of 0,122 can be interpreted as a 12,2% certainty about the outcome given by all of the predictors simultaneously. This is less than 25%, and it means, that the model is a poor predictive model. The analysis of variance (ANOVA) table produced a F (Fisher) statistic of 4236 compatible with a p-value of <0,001. Overall the poor predictive model yet produced a result significantly different from a 0% certainty at a p-value at <0,001. The coefficients table shows which of the predictor variables are strong predictors. Two of the eight of the predictors, Buttered toast and Jelly donut, were statistically significant at a $p < 0,05$ level. Instead of a traditional linear regression, a multinomial logistic regression may provide a better fit model for the data, that are categorical rather than continuous in nature.

Command:

Menu....Analyze....Regression....Multinomial Logistic Regression....Dependent: Menu scenarios....Independent(s): enter all of the predictor variables....click OK.

The underneath tables are in the output sheets.

Parameter Estimates

Menu scenarios[a]		B	Std. Error	Wald	df	Sig.	Exp(B)	95% Confidence Interval for Exp (B)	
								Lower Bound	Upper Bound
Overall preference	Intercept	-2,440	2,855	,730	1	,393			
	TP	,078	,061	1,632	1	,201	1,081	,959	1,219
	BT	-,123	,091	1,839	1	,175	,884	,739	1,057
	JD	,151	,070	4,611	1	,032	1,163	1,013	1,334
	TMd	-,050	,073	,466	1	,495	,952	,825	1,097
	TMn	,099	,094	1,112	1	,292	1,104	,918	1,328
	CB	-,043	,082	,280	1	,597	,958	,816	1,124
	DP	,020	,074	,073	1	,788	1,020	,882	1,180
	CMB	,121	,080	2,274	1	,132	1,129	,964	1,322
Breakfast, with juice, bacon and eggs, and beverage	Intercept	2,963	3,443	,740	1	,390			
	TP	,095	,075	1,638	1	,201	1,100	,951	1,273
	BT	-,329	,112	8,638	1	,003	,720	,578	,896
	JD	,095	,089	1,150	1	,284	1,100	,924	1,309
	TMd	-,217	,081	7,236	1	,007	,805	,688	,943
	TMn	-,093	,104	,794	1	,373	,911	,743	1,118
	CB	-,136	,101	1,805	1	,179	,873	,715	1,065
	DP	,114	,083	1,864	1	,172	1,120	,952	1,319
	CMB	-,020	,092	,045	1	,832	,981	,819	1,174
Breakfast, with juice, cold cereal, and beverage	Intercept	2,087	3,017	,478	1	,489			
	TP	,034	,064	,280	1	,597	1,034	,913	1,172
	BT	-,121	,097	1,554	1	,213	,886	,732	1,072
	JD	,094	,075	1,589	1	,208	1,099	,949	1,273
	TMd	-,194	,074	6,810	1	,009	,824	,712	,953
	TMn	-,062	,099	,394	1	,530	,940	,775	1,140
	CB	-,093	,087	1,137	1	,286	,911	,768	1,081
	DP	,074	,073	1,026	1	,311	1,077	,933	1,243
	CMB	,021	,083	,063	1	,802	1,021	,868	1,201
Breakfast, with juice, pancakes, sausage, and beverage	Intercept	5,397	3,158	2,921	1	,087			
	TP	,039	,066	,338	1	,561	1,039	,913	1,183
	BT	-,208	,098	4,540	1	,033	,812	,671	,983
	JD	-,021	,077	,075	1	,785	,979	,841	1,139
	TMd	-,220	,076	8,426	1	,004	,802	,691	,931
	TMn	-,123	,099	1,546	1	,214	,884	,729	1,073
	CB	-,149	,090	2,778	1	,096	,861	,723	1,027
	DP	,073	,075	,955	1	,328	1,076	,929	1,247
	CMB	-,054	,086	,395	1	,530	,948	,801	1,121
Breakfast, with beverage only	Intercept	1,773	2,894	,375	1	,540			
	TP	,008	,057	,020	1	,888	1,008	,901	1,128
	BT	,006	,102	,004	1	,949	1,007	,824	1,229
	JD	,105	,071	2,213	1	,137	1,111	,967	1,277
	TMd	-,085	,073	1,366	1	,242	,919	,797	1,059
	TMn	-,126	,100	1,577	1	,209	,882	,725	1,073
	CB	-,099	,083	1,428	1	,232	,906	,770	1,066
	DP	-,057	,077	,562	1	,453	,944	,813	1,097
	CMB	,030	,081	,135	1	,713	1,030	,879	1,208

a. The reference category is: Snack, with beverage only.

Pseudo R-Square

Cox and Snell	,400
Nagelkerke	,411
McFadden	,142

Various predictor variables were significant predictors of different outcome values, the menu scenarios. However, for comparisons with kernel ridge regression we are particularly interested in the overall R square values, giving the percentage of certainty of the predictors about the outcome. However, traditional R square values are not provided by the multinomial model. Fortunately, pseudo R square values based on log likelihood statistics are provided instead. Cox and Snell pseudo R squares are commonly used. The value of 0,400 is a lot better fit than is the R square of 0,112 from the traditional linear regression of the data.

We will perform kernel ridge regressions next, in order to find out, whether a still better data fit is possible.

23.5 Kernel Ridge Regressions

Subsequently, various kernel ridge regressions were performed of the same data file.

Command:

Analyze....Regression....Kernel Ridge Regression....Dependent: menu scenarios (srcid)....Independent(s): enter all of the eight predictor variables....click Linear.... click OK.

Model Summary[a,b]

Kernel	Alpha	R Square
Linear	1,000	,066

a. Dependent Variable: srcid

b. Model: TP, BT, JD, TMd, TMn, CB, DP, CMB

The above result is not any better than the result from the traditional linear regression, R square 0,066 vs 0,122. We will try and assess more kernel density models.

Command:

Analyze....Regression....Kernel Ridge Regression....Dependent: menu scenarios (srcid)....Independent(s): enter all of the eight predictor variables....click Linear.... click Additive_chi2....click OK.

Model Summary[a,b]

Kernel	Alpha	R Square
Additive_chi2	1,000	,182

a. Dependent Variable: srcid

b. Model: TP, BT, JD, TMd, TMn,
   CB, DP, CMB

The above output result is slightly better, but R squares over 0,50 are very welcome. Therefore, more kernel density models will be assessed using similar demands.

Model Summary[a,b]

Kernel	Alpha	Gamma	R Square
Chi2	1,000	1,000	,143

a. Dependent Variable: srcid

b. Model: TP, BT, JD, TMd, TMn, CB, DP,
   CMB

Model Summary[a,b]

Kernel	Alpha	R Square
Cosine	1,000	,102

a. Dependent Variable:
   srcid

b. Model: TP, BT, JD, TMd,
   TMn, CB, DP, CMB

Model Summary[a,b]

Kernel	Alpha	Gamma	R Square
Laplacian	1,000	,125	,510

a. Dependent Variable: srcid

b. Model: TP, BT, JD, TMd, TMn, CB, DP, CMB

Model Summary[a,b]

Kernel	Alpha	Gamma	Coef0	Degree	R Square
Polynomial	1,000	,125	1,000	3,000	,659

a. Dependent Variable: srcid

b. Model: TP, BT, JD, TMd, TMn, CB, DP, CMB

Model Summary[a,b]

Kernel	Alpha	Gamma	R Square
RBF	1,000	,125	-,019

a. Dependent Variable: srcid

b. Model: TP, BT, JD, TMd, TMn, CB, DP, CMB

Model Summary[a,b]

Kernel	Alpha	Gamma	Coef0	R Square
Sigmoid	1,000	,125	1,000	,000

a. Dependent Variable: srcid

b. Model: TP, BT, JD, TMd, TMn, CB, DP, CMB

The best datafits were provided by the Laplacian and the Polynomial kernel density models (respectively R square values 0,510 and 0,659, meaning 51,0% and 65,9% predictive certainties). The traditional multiple variables linear regression provided an R square value of only 0,122 (12,2% predictive certainty).

Note, that the RBF (radial basis function) kernel density model produced a negative R square. Explanation is given in the Chap. 5. "Some Terminology". The above kernel ridge regressions used eight different kernel density models. Graphical presentations of all of them are also in the Chap. 5 entitled "Some Terminology".

23.6 Conclusion

23.6.1 Summaries of the Traditional Multiple Variables Linear Regression

The R square value of 0,122 can be interpreted as a 12,2% certainty about the outcome given by all of the predictors simultaneously. This is less than 25%, and it means, that the model is a poor predictive model. The analysis of variance (ANOVA) table produced a F (Fisher) statistic of 4236 compatible with a p-value of <0,001. Overall the poor predictive model still produced a result significantly different from a 0% certainty at a p-value at <0,001. The coefficients table shows which of the predictor variables are strong predictors. Two of the eight of the predictors, Buttered toast and Jelly donut, were statistically significant at a $p < 0,05$ level. We also performed multinomial logistic regressions to find out whether a better data fit is possible. The overall (pseudo) R square rose from 0,122 to 0, 400.

23.6.2 Summaries of the Kernel Ridge Regressions

The best data fits were provided by the Laplacian and the Polynomial kernel density models (respectively R square values 0,510 and 0,659, meaning 51,0% and 65,9% predictive certainties). The traditional multiple variables linear regression provided an R square value of only 0,122 (12,2% predictive certainty).

Note, that the RBF (radial basis function) kernel density model produced a negative R square. Explanation is given in the Chap. 5. entitled "Some Terminology",

23.7 References

All of the chapters of the current edition start with a brief review of the traditional analytic method of the different regression methods prior to the review of the relevant kernel ridge regression method. For the purpose, generally, data examples are used from the recent edition "Regression Analyses in Clinical Research for Starters and 2nd Levelers 2nd Edition, Springer Heidelberg Germany 2021", by the same authors. For a better understanding of differences between traditional and kernel ridge regressions, readers may benefit from the study of this edition first.

To readers requesting still more background, theoretical and mathematical information of computations given, several textbooks complementary to the current production and written by the same authors are available: Statistics applied to

clinical studies 5th edition, 2012, Machine learning in medicine a complete overview 2nd edition, 2020, SPSS for starters and 2nd levelers 2nd edition, 2015, Clinical data analysis on a pocket calculator 2nd edition, 2016, Understanding clinical data analysis from published research, 2016, all of them edited by Springer Heidelberg Germany.

Chapter 24
Effect of Personal Factors on Anorexia, 217 Persons, Traditional Linear Regression vs Kernel Ridge Regressions

Abstract The effect of personal factors on anorexia was studied in 217 persons. The R square value of the traditional multiple variables linear regression of 0,481 can be interpreted as a 48,1% certainty about the outcome given by all of the personal factors simultaneously. This is close to 50%, which is a reasonable, but not yet strong prediction level. With kernel ridge regressions the best fit model was provided by the Polynomial kernel density model (R square 0,745), and this was much better than the traditional linear regression R square value of 0,481. More very good kernel density models were observed:

the RBF kernel density model: R square 0,625
the Laplacian kernel density model: R square 0,620
the Chi2 kernel density model: R square 0,663
the Additive_chi2 kernel density model: R square 0,509.

All of them were better fit than the traditional linear regression with an R square value of 0,481.

Keywords Traditional multiple variables linear regression · Polynomial kernel density · Chi2 kernel density

24.1 Summary

The effect of personal factors on anorexia was studied in 217 persons.

Supplementary Information The online version contains supplementary material available at [https://doi.org/10.1007/978-3-031-10717-7_24].

24.1.1   Summaries of the Traditional Multiple Variables
Linear Regression

The R square value of 0,481 can be interpreted as a 48,1% certainty about the outcome given by all of the personal factors simultaneously. This is close to 50%, and it means that the model is a reasonable, but not yet strong prediction model. The analysis of variance (ANOVA) table produced a F (Fisher) statistic of 17,278 compatible with a p-value of <0,001. The coefficients table showed which of the predictor variables were the strongest predictors. Six of the 11 predictors were statistically significant at a $p < 0,10$ level.

24.1.2   Summaries of the Kernel Ridge Regression

The best fit model was provided by the Polynomial kernel density model (R square 0,745), and this was much better than the traditional linear regression R square value of 0,481.

More very good kernel density models were observed. For example:

the RBF kernel density model: R square 0,625
the Laplacian kernel density model: R square 0,620
the Chi2 kernel density model: R square 0,663
the Additive_chi2 kernel density model: R square 0,509.

All of them were better fit than the traditional linear regression with an R square value of 0,481.

24.2   Introduction

The Data file entitled "anorexia" is in SpringerLink supplementary files. This data file was partly taken from the "Samples site" of IBM SPSS Statistics version 28 0.1.0. In 217 persons the effect of 11 personal factors on anorexia was studied. Anorexia was estimated as body weight scores. First, a traditional multiple variables linear regression was performed. Second, kernel ridge regressions were done using eight different kernel density models. The traditional multiple variables linear regression provided a p-value of <0,001, R square 0,481. The kernel ridge regression with linear kernel density model provided an R square of 0,477.

This was virtually the same as that of the traditional linear regression. However, more kernel density models are available, and some of them provided much better R square values, sometimes even as large as large as 0,745. And so, kernel ridge regression is able to provide much better data fit than traditional linear regression does.

24.3   Data Example

The Data file entitled "anorexia" is in SpringerLink supplementary files. This data file was partly taken from the Samples site of IBM SPSS Statistics version 28 0.1.0. In 217 persons the effect of 11 personal factors on anorexia was studied. Anorexia was estimated as body weight scores. First, a traditional multiple variables linear regression was performed. Second, kernel ridge regressions were done using eight different kernel density models. The first 12 patients of the data file is underneath. The remainder is of course in the supplementary files.

Variables 1 - 12

1	2	3	4	5	6	7	8	9	10	11	12
1	1	1	4	4	1	3	2	2	1	1	1
1	1	1	4	4	2	3	2	2	1	1	1
1	1	1	4	4	3	3	2	1	2	1	1
1	1	1	4	4	2	3	2	2	3	1	1
3	1	1	4	4	2	1	2	2	3	1	1
1	1	1	4	4	2	1	2	2	3	1	1
1	1	1	4	4	2	1	1	1	1	1	1
1	1	1	4	4	3	3	2	2	3	2	1
1	1	1	4	4	2	3	3	2	1	1	1
1	1	1	4	4	2	3	2	2	1	1	1
1	1	1	4	4	1	2	1	1	3	1	1
1	1	1	4	4	1	2	2	2	3	1	1

Variable 1 = body weight score
The variables 2-12 are respectively given underneath.

| Menstruation |
| Restriction of food intake (fasting) |
| Binge eating |
| Vomiting |
| Hyperactivity |
| Friends |
| School/employment record |
| Sexual behavior |
| Mental state (mood) |
| Preoccupation with food and weight |
| Patient Diagnosis |

24.4   Traditional Linear Regression

First, a traditional linear regression with 11 predictors and body weight scores as outcome was performed. Download the 217 person data file in your computer installed with SPSS statistical software 2022 version 28 0.1.0. Start by opening the data file.

Command:

Analyze....Menu....Regression....Linear Regression....Dependent: enter body weight score....Independent (s): enter all of the 11 predictors....click OK.

   In the output sheets are the underneath tables.

Model Summary

Model	R	R Square	Adjusted R Square	Std. Error of the Estimate
1	,694[a]	,481	,453	,989

a. Predictors: (Constant), Patient Diagnosis, School/employment record, Vomiting, Mental state (mood), Friends, Binge eating, Hyperactivity, Restriction of food intake (fasting), Sexual behavior, Menstruation, Preoccupation with food and weight

ANOVA[a]

Model		Sum of Squares	df	Mean Square	F	Sig.
1	Regression	185,773	11	16,888	17,278	<,001[b]
	Residual	200,383	205	,977		
	Total	386,157	216			

a. Dependent Variable: Body Weight

b. Predictors: (Constant), Patient Diagnosis, School/employment record, Vomiting, Mental state (mood), Friends, Binge eating, Hyperactivity, Restriction of food intake (fasting), Sexual behavior, Menstruation, Preoccupation with food and weight

Coefficients[a]

Model		Unstandardized Coefficients		Standardized Coefficients	t	Sig.
		B	Std. Error	Beta		
1	(Constant)	-,613	,487		-1,259	,210
	Menstruation	,111	,097	,075	1,143	,254
	Restriction of food intake (fasting)	,355	,077	,319	4,604	<,001
	Binge eating	-,160	,068	-,130	-2,334	,021
	Vomiting	-,083	,089	-,053	-,935	,351
	Hyperactivity	,386	,118	,192	3,263	,001
	Friends	-,138	,099	-,078	-1,398	,164
	School/employment record	,333	,097	,202	3,428	<,001
	Sexual behavior	,171	,132	,084	1,295	,197
	Mental state (mood)	,198	,113	,102	1,746	,082
	Preoccupation with food and weight	,177	,127	,098	1,393	,165
	Patient Diagnosis	,150	,068	,124	2,206	,028

a. Dependent Variable: Body Weight

The R square value of 0,481 can be interpreted as a 48,1% certainty about the outcome given by all of the predictors simultaneously. This is close to 50%, and it means that the model is a reasonable, though not a strong prediction model.

The analysis of variance (ANOVA) table produced an F (Fisher) statistic of 17,278 compatible with a p-value of <0,001.

The coefficients table shows which of the predictor variables are the strongest predictors. Six of the 11 predictors were statistically significant at a $p < 0,10$ level.

We will perform kernel ridge regressions to find out whether a better data fit is possible.

24.5 Kernel Ridge Regressions

Subsequently, various kernel ridge regressions were performed of the same data file.

Command:

Analyze....Regression....Kernel Ridge Regression....Dependent: weight....Independent(s): enter all of the 11 predictor variables....click Linear....click OK.

Model Summary[a,b]

Kernel	Alpha	R Square
Linear	1,000	,477

a. Dependent Variable: weight

b. Model: mens, fast, binge, vomit, hyper, frie, school, sbeh, mood, preo, diag

The above table is in the output. The kernel density model with a Q square value of 0,477 is OK, but more sensitivity of testing is welcome. Therefore, more kernel density models will be assessed.

Command:

Analyze....Regression....Kernel Ridge Regression....Dependent: weight....Independent(s): enter all of the 11 predictor variables....click Linear....click Additive_chi2....click OK.

Model Summary[a,b]

Kernel	Alpha	R Square
Additive_chi2	1,000	,509

a. Dependent Variable: weight

b. Model: mens, fast, binge, vomit, hyper, frie, school, sbeh, mood, preo, diag

The Additive_chi2 kernel density model fits the data slightly better fit than the linear kernel density model does, but better models are still welcome. More models are assessed giving again similar commands.

Model Summary[a,b]

Kernel	Alpha	Gamma	R Square
Chi2	1,000	1,000	,663

a. Dependent Variable: weight

b. Model: mens, fast, binge, vomit,
   hyper, frie, school, sbeh, mood, preo,
   diag

Model Summary[a,b]

Kernel	Alpha	R Square
Cosine	1,000	,357

a. Dependent Variable:
   weight

b. Model: mens, fast, binge,
   vomit, hyper, frie, school,
   sbeh, mood, preo, diag

Model Summary[a,b]

Kernel	Alpha	Gamma	R Square
Laplacian	1,000	,091	,620

a. Dependent Variable: weight

b. Model: mens, fast, binge, vomit, hyper,
   frie, school, sbeh, mood, preo, diag

Model Summary[a,b]

Kernel	Alpha	Gamma	Coef0	Degree	R Square
Polynomial	1,000	,091	1,000	3,000	,745

a. Dependent Variable: weight

b. Model: mens, fast, binge, vomit, hyper, frie, school, sbeh, mood,
   preo, diag

Model Summary[a,b]

Kernel	Alpha	Gamma	R Square
RBF	1,000	,091	,625

a. Dependent Variable: weight

b. Model: mens, fast, binge, vomit,
hyper, frie, school, sbeh, mood, preo,
diag

Model Summary[a,b]

Kernel	Alpha	Gamma	Coef0	R Square
Sigmoid	1,000	,091	1,000	-,004

a. Dependent Variable: weight

b. Model: mens, fast, binge, vomit, hyper, frie, school,
sbeh, mood, preo, diag

The best fit model was provided by the Polynomial kernel density model (R square 0,745), and this was much better than the traditional linear regression R square value of 0,481. More very good kernel density models were observed. For example:

the RBF (radial basis function) kernel density model   R square 0,625
the Laplacian kernel density model: R square 0,620
the Chi2 kernel density model: R square 0,663
the Additive_chi2 kernel density model: R square 0,509.

All of them were better fit than the traditional linear regression was with an R square value of 0,481. Note, that the above kernel ridge regressions used eight different kernel density models. Graphical presentations of all of them are in the Chap. 5, entitled "Some Terminology".

24.6 Conclusions

24.6.1 Summaries of the Traditional Multiple Variables Linear Regression

The R square value of 0,481 can be interpreted as a 48,1% certainty about the outcome given by all of the personal factors simultaneously. This is close to 50%, and it means that the model is a reasonable, but not yet a strong prediction model.

The analysis of variance (ANOVA) table produced a F (Fisher) statistic of 17,278 compatible with a p-value of <0,001.

The coefficients table shows which of the predictor variables are the strongest predictors. Six of the 11 predictors were statistically significant at a $p < 0,10$ level.

24.6.2 Summaries of the Kernel Ridge Regression

The best fit model was provided by the Polynomial kernel density model (R square 0,745, 74,5% certainty), and this was much better than the traditional linear regression R square value of 0,481.

More very good kernel density models were observed. For example:

the RBF kernel density model: R square 0,625
the Laplacian kernel density model: R square 0,620
the Chi2 kernel density model: R square 0,663
the Additive_chi2 kernel density model: R square 0,509.

All of them were better fit than the traditional linear regression with an R square value of 0,481.

24.7 References

All of the chapters of the current edition start with a brief review of the traditional analytic method of the different regression methods prior to the review of the relevant kernel ridge regression method. For the purpose, generally, data examples are used from the recent edition "Regression Analyses in Clinical Research for Starters and 2nd Levelers 2nd Edition, Springer Heidelberg Germany 2021", by the same authors. For a better understanding of differences between traditional and kernel ridge regressions, readers may benefit from the study of this edition first.

To readers requesting still more background, theoretical and mathematical information of computations given, several textbooks complementary to the current production and written by the same authors are available: Statistics applied to clinical studies 5th edition, 2012, Machine learning in medicine a complete overview 2nd edition, 2020, SPSS for starters and 2nd levelers 2nd edition, 2015, Clinical data analysis on a pocket calculator 2nd edition, 2016, Understanding clinical data analysis from published research, 2016, all of them edited by Springer Heidelberg Germany.

Chapter 25
Effect on Weight Loss of Physical Exercise, Calorie Intake, Their Interaction, and Age, in 64 Patients, Traditional Regression vs Kernel Ridge Regression

Abstract The effect on weight loss of physical exercise, calorieintake, and age was studied in 64 patients. The R square value of the traditional multiple variables linear regression of 0,502 can be interpreted as a 50,2% certainty about the outcome given by all of the predictors simultaneously. The coefficients table shows which of the predictor variables are the strongest predictors. Calorieintake and interaction between physical exercise and interaction between physical exercise and calorieintake are the strongest predictors and significantly different from zero correlations at p-values of <0,000 and < 0,002. With kcrnel ridge regressions better results were obtained. For example R square = 0,641 for the polynomial kernel density model and 0, 677 for the Additive_chi2 kernel density model, and so kernel density provided better data fit models than did traditional linear regression.

Keywords Traditional multiple variables linear regression · Coefficients table · Kernel ridge regression · Polynomial kernel density · Additive_chi2 kernel density

25.1 Summary

In this chapter the effect on weight loss of physical exercise, calorieintake, and age was studied in 64 patients. The outcome of clinical research is, generally, affected by many more factors than a single one, and multiple variables regression assumes, that these factors act independently of one another, but why should they not affect one another. With multicollinearity one or more x-variables have a strong correlation with one another. How does ridge regression deal with multicollinearity? Ridge Regression is a technique for analyzing multiple regression data that suffer from multicollinearity. **By adding a degree of bias to the regression estimates**, ridge regression reduces the standard errors. It is then hoped, that the net effect will be more adequate.

Supplementary Information The online version contains supplementary material available at [https://doi.org/10.1007/978-3-031-10717-7_25].

25.1.1 Summaries of the Traditional Multiple Variables Linear Regression

The R square value of 0,502 can be interpreted as a 50,2% certainty about the outcome given by all of the predictors simultaneously. This is close to 50%, and it means that the model is a reasonable, though not a strong prediction model. The analysis of variance (ANOVA) table produced an F (Fisher) statistic of 14,890 compatible with a p-value of <0,000. Obviously, weight loss is significantly dependent on the combination of all of the x-variables.

The coefficients table shows which of the predictor variables are the strongest predictors. Calorieintake and interaction between physical exercise and interaction between physical exercise and calorieintake are the strongest predictors and significantly different from zero correlations at p-values of <0,000 and < 0,002. In order to find out whether kernel ridge regression is able to provide a still better predictive score for this data, we will, subsequently, perform kernel ridge regressions.

25.1.2 Summaries of Kernel Ridge Regressions

The R square values of some kernel density models were considerably better than the R square of the traditional linear regression (0,502). For example, 0,641 for the polynomial kernel density model and 0, 677 for the Additive_chi2 kernel density model, and so kernel density provided better data fit models than did traditional linear regression.

25.2 Introduction

The outcome of clinical research is, generally, affected by many more factors than a single one, and multiple variables regression assumes, that these factors act independently of one another, but why should they not affect one another. With multicollinearity one or more x-variables have a strong correlation with one another. It can be detected by a correlation matrix. If one by one correlations have a regression coefficient over 85%, then multicollinearity is in the data. Also a sharp increase in a t-value for the coefficient of an x-variable when another x-variable is removed from the model suggests the presence of multicollinearity. Multicollinearity does not reduce the predictive power or reliability of the model as a whole, at least within the sample data set; it only affects calculations regarding individual predictors. How does ridge regression deal with multicollinearity? Ridge Regression is a technique for analyzing multiple regression data that suffer from multicollinearity. **By adding a degree of bias to the regression estimates**, ridge regression reduces the standard errors. It is then hoped, that the net effect will be to give estimates that are more reliable.

25.3 Data Example

In a 64 patient study the effect on weight loss of three predictors, including physical exercise, calorieintake, and their interaction was assessed. The datafile was entitled "interaction", and can be downloaded from SpringerLink supplementary files. For analysis SPSS statistical software 2022 version 28 0.1.0 must be mounted on your computer. The first 9 patients are in the underneath table.

Variables

1	2	3	4	5
1,00	,00	3000,00	,00	64,00
28,00	6,00	3000,00	18000,00	34,00
27,00	6,00	3000,00	18000,00	25,00
30,00	6,00	3000,00	18000,00	34,00
27,00	6,00	1000,00	6000,00	45,00
29,00	,00	2000,00	,00	52,00
31,00	3,00	2000,00	6000,00	59,00
30,00	3,00	1000,00	3000,00	58,00
29,00	3,00	1000,00	3000,00	47,00

1. weightloss
2. physical exercise
3. calorieintake
4. interaction between physical exercise and calorie intake
5. age

25.4 Traditional Linear Regression

First a traditional multiple variables linear regression will be performed. SPSS statistical software must be installed in your computer. Start by downloading the data file of the example entitled "interaction".

Command:

Analyze....Regression....Linear....Dependent: weightloss.... Independent(s): physical exercise, calorieintake, interaction calorieintake x exercise, age....click OK.
   The underneath tables are in the output sheets.

Model Summary

Model	R	R Square	Adjusted R Square	Std. Error of the Estimate
1	,709[a]	,502	,469	7,69754

a. Predictors: (Constant), age, calorieintake, exercise, interaction

ANOVA[b]

Model		Sum of Squares	df	Mean Square	F	Sig.
1	Regression	3529,063	4	882,266	14,890	,000[a]
	Residual	3495,875	59	59,252		
	Total	7024,938	63			

a. Predictors: (Constant), age, calorieintake, exercise, interaction
b. Dependent Variable: weightloss

Coefficients[a]

Model		Unstandardized Coefficients		Standardized Coefficients	t	Sig.
		B	Std. Error	Beta		
1	(Constant)	31,400	5,653		5,555	,000
	exercise	-,199	,974	-,048	-,204	,839
	calorieintake	-,010	,002	-,826	-6,210	,000
	interaction	,001	,000	,898	3,218	,002
	age	,057	,099	,063	,577	,566

a. Dependent Variable: weightloss

The R square value of 0,502 can be interpreted as a 50,2% certainty about the outcome given by all of the predictors simultaneously. This is close to 50%, and it means that the model is a reasonable, though not a strong prediction model.

The analysis of variance (ANOVA) table produced an F (Fisher) statistic of 14,890 compatible with a p-value of <0,000. Obviously, weightloss is significantly dependent on the combination of all of the x-variables.

The coefficients table shows which of the predictor variables are the strongest predictors. Calorieintake and interaction between physical exercise and interaction between physical exercise and calorieintake are the strongest predictors and significantly different from zero correlations at p-values of <0,000 and < 0,002.

In order to find out whether kernel ridge regression is able to provide a still better predictive score for this data, we will, subsequently, perform kernel ridge regressions.

25.5 Kernel Ridge Regression

Subsequently, various kernel ridge regressions were performed of the.
  same datafile.

Command:

Analyze....Regression....Kernel Ridge Regression....Dependent: weightloss....Independent(s): enter all of the 4 predictor variables....click Linear....click OK.

Model Summary[a,b]

Kernel	Alpha	R Square
Linear	1,000	,242

a. Dependent Variable: weightloss

b. Model: exercise, calorie, interaction, age

The above table is in the output. The kernel density model with a Q square value of 0,242 (24,2% certainty prediction) is pretty poor, better sensitivity of testing is welcome. Therefore, more kernel density models will be assessed.

Command:

Analyze....Regression....Kernel Ridge Regression....Dependent: weight....Independent(s): enter all of the 11 predictor variables....click Linear....click Additive_chi2....click OK.

Model Summary[a,b]

Kernel	Alpha	R Square
Additive_chi2	1,000	,677

a. Dependent Variable:
   weightloss

b. Model: exercise, calorie,
   interaction, age

The above Additive_chi2 kernel density model summary table gives a much better datafit with 67,7% predictive certainty. Similar commands are given for additional kernel density models.

Model Summary[a,b]

Kernel	Alpha	Gamma	R Square
Chi2	1,000	1,000	,384

a. Dependent Variable: weightloss

b. Model: exercise, calorie, interaction,
   age

Model Summary[a,b]

Kernel	Alpha	Gamma	R Square
Laplacian	1,000	,250	,253

a. Dependent Variable: weightloss

b. Model: exercise, calorie, interaction, age

Model Summary[a,b]

Kernel	Alpha	R Square
Cosine	1,000	,221

a. Dependent Variable:
   weightloss

b. Model: exercise, calorie,
   interaction, age

Model Summary[a,b]

Kernel	Alpha	Gamma	Coef0	Degree	R Square
Polynomial	1,000	,250	1,000	3,000	,641

a. Dependent Variable: weightloss

b. Model: exercise, calorie, interaction, age

Model Summary[a,b]

Kernel	Alpha	Gamma	R Square
RBF	1,000	,250	-,018

a. Dependent Variable: weightloss

b. Model: exercise, calorie, interaction, age

For explanation of the negative R square value with the RBF (radial basis function) kernel density model see Chap. 5 "Some Terminology".

Model Summary[a,b]

Kernel	Alpha	Gamma	Coef0	R Square
Sigmoid	1,000	,250	1,000	-,001

a. Dependent Variable: weightloss

b. Model: exercise, calorie, interaction, age

The R square values of some kernel density models were considerably better than the R square of the traditional linear regression (0,502). For example, 0,641 for the Polynomial kernel density model, and 0, 677 for the Additive_chi2 kernel density model, and so kernel density provided better data fit models than did traditional linear regression.

The above eight kernel ridge regressions used eight different kernel density models. All of them are graphically presented in the Chap. 5, entitled "Some Terminology". Sometimes with kernel ridge regressions R square values can be negative. This indicates poor data fit. See also the Chap. 5, entitled "Some Terminology".

25.6   Conclusion

With multicollinearity one or more x-variables have a strong correlation with one another. How does ridge regression deal with multicollinearity? Ridge Regression is a technique for analyzing multiple regression data that suffer from multicollinearity. **By adding a degree of bias to the regression estimates**, ridge regression reduces the standard errors. It is then hoped, that the net effect will be more adequate.

25.6.1   Summaries of the Traditional Multiple Variables Linear Regression

The R square value of 0,502 can be interpreted as a 50,2% certainty about the outcome given by all of the predictors simultaneously. This is over 50%, and it means that the model's predictive potential is a reasonable, though not very strong prediction. The analysis of variance (ANOVA) table produced an F (Fisher) statistic of 14,890 compatible with a p-value of <0,000. Obviously, weight loss is significantly dependent on the combination of all of the x-variables.

The coefficients table shows which of the predictor variables are the strongest predictors. Calorieintake and interaction between physical exercise and interaction between physical exercise and calorieintake are the strongest predictors and significantly different from zero correlations at p-values of <0,000 and < 0,002. In order to find out whether kernel ridge regression is able to provide a still better predictive score for this data, we will, subsequently, perform kernel ridge regressions.

25.6.2   Summaries of Kernel Ridge Regressions

The R square values of some kernel density models were considerably better than the R square of the traditional linear regression (0,502). For example, 0,641 for the polynomial kernel density model and 0, 677 for the Additive_chi2 kernel density model, and so kernel density provided better data fit models than did traditional linear regression.

25.7   References

All of the chapters of the current edition start with a brief review of the traditional analytic method of the different regression methods prior to the review of the relevant kernel ridge regression method. For the purpose, generally, data examples are used from the recent edition "Regression Analyses in Clinical Research for

Starters and 2nd Levelers 2nd Edition, Springer Heidelberg Germany 2021", by the same authors. For a better understanding of differences between traditional and kernel regressions, readers may benefit from the study of this edition first.

To readers requesting still more background, theoretical and mathematical information of computations given, several textbooks complementary to the current production and written by the same authors are available: Statistics applied to clinical studies 5th edition, 2012, Machine learning in medicine a complete overview, 2015, SPSS for starters and 2nd levelers 2nd edition, 2015, Clinical data analysis on a pocket calculator 2nd edition, 2016, Understanding clinical data analysis from published research, 2016, all of them edited by Springer Heidelberg Germany.

Chapter 26
Summaries

Abstract The main results of 20 clinical data studies are summarized. Analyses consisted of traditional statistical regressions as well as the novel kernel ridge regressions.

Keywords Clinical data studies · Traditional regressions · Kernel ridge regressions

26.1 Chapter 1

With traditional regression methods, the outcome values are assumed to be normally distributed around the regression line/curve. With non-normal outcome data, that remain non-normal in spite of transformations (Likert scales is a notorious example), data distributions may be skewed, and nonparametric regression analysis may provide better data fit than traditional parametric models do. Methods including nonparametric regression are pretty new, and not yet widely applied. They include: kriging, otherwise called Gaussian process regression, decision trees, and bagged (bootstrap aggregated) regression trees, kernel regressions, and median regression, otherwise called robust regression. The AICs (Akaike Information Criterion goodness of fit tests) of the full loglikelihood model and the kernel model were respectively 1085 and 920, difference 165. This means, that the probability of the kernel regression model to minimize information loss is $e^{(165)/2}$ times less worse than the full loglikelihood model is. Obviously, the kernel model performed endlessly better. Kernel regression provided a stronger goodness of fit than did traditional loglikelihood testing. However, no p-values are obtained. Traditional regression analysis produced two strong predictors of body surface, namely weight and height, and a borderline significant age effect of $p = 0,040$.

26.2 Chapter 2

Three problems with kernel regression have to be accounted, and have, indeed, have been accounted for by the recent kernel ridge methodology.

1. Kernel regression is more sensitive than traditional ordinary least squares regression, but as shown in the previous chapter it is a *discretization model*. By the add-up sum of Gaussians, continuous variables are converted into discrete ones, otherwise discretized ones.
2. Another problem is that of increasing mathematical complexity with multidimensional data. However, the *kernel trick*, which will be explained in the next few lines, offers a wonderful solution to the problem.
3. A third problem, is that of *overfitting*, i.e., data patterns wider than compatible with random sampling. It can, however, be corrected with some kind of data regularization, where regression coefficients (b-values) are penalized to a lower level. Ridge, lasso, elastic net regularization are accepted manners. And with kernel regressions generally ridge regularization can be successfully used. It reduces b-values according to $b_{ridge} = b/(1 + \lambda)$ where λ = shrinking factor.

26.3 Chapter 3

The effect of linear predictor on continuous outcome was assessed using different scales of the predictor values.

26.3.1 Summaries of the Traditional Linear Regressions

In summary the traditional R square values of the scale 1–3 models were respectively

scale1 R square = 0,277 = 27,7% certainty about the outcome
scale2 R square = 0,281 = 28,1% certainty about the outcome
scale3 R square = 0,380 = 38,0% certainty about the outcome

The p-values of the traditional linear regressions of the three scale models were respectively

scale1 p-value = 0,079
scale2 p-value = 0,076
scale3 p-value = 0,033.

26.3.2 Summaries of the Kernel Ridge Regressions

The **scale 1** kernel ridge regressions produced four kernel density models with R squares over 0,250. The best fit R square was provided by the polynomial kernel density model, 0,455.

The **scale 2** kernel ridge regressions produced five kernel density models with R squares over 0,250. The best fit R square was provided by the polynomial kernel density model, 0,808.

The **scale 3** kernel ridge regressions produced six kernel density models with R squares over 0,250. The best fit R square was provided by the polynomial kernel density model, 0,962.

26.3.3   In Conclusion

Optimal scaling is a possibility to improve the correlation between predictors and outcomes. KRR (kernel ridge regression) provides optimally fit correlations, and performs even better than optimal scaling for the purpose of optimized predictive modeling.

26.4   Chapter 4

The theory of kernel ridge regressions was first introduced by AE Hoerl and RW Kennard, computer scientists from the University of Delaware. The authors' 1970 seminal paper in Technometrics (1970; 8: 27–51) was entitled "Ridge regression based estimation of nonorthogonal problems".

Theoretical advantages include (1) kernel trick for reduced arithmetic complexity, (2) estimation of uncertainty through Gaussians instead of histograms, (3) corrected data-overfit through ridge regularization, (4) availability of multiple alternative kernel density models for improved data fit.

A brief search of competitive kernel ridge regression publications revealed

one genomic study,
one facial surgery study,
one genetic study.

The remainder of publications involved chemistry studies, econometry studies, and studies from basic sciences like nature, biology and physics. Also climate studies were observed.

26.5   Chapter 5

Some Terminology:

Exhausive searching
Graphical frequency distributions
Histograms and graphical frequency distributions vs mathematical frequency
    distributions
Interpretation of R Square values
IoT (internet of things)
Kernel density model
Kernel trick
Lacking p-values
Linear kernel model
Mathematical frequency distributions
Misnomer
Multicollinearity
Multivariate kernel ridge regression
Overfitting
Problems with R Square values
Pseudo Re Square values
R Square values that, with a kernel ridge regression, can be negative
Ridge Regularization
Scikit-learn or Sklearn
Study Stream
Stream Study
Test-Retest Reliability
Final note

26.6   Chapter 6

In observational research event rates are often very much age and sex dependent and
a model routinely adjusting these confounders is welcome. A data example will be
given from Kirkwood and Sterne (Standardization, in: Medical Statistics, Chap. 25,
Blackwell Science, Oxford UK 2003). The authors studied in 12,816 onchocerciasis
patientyears, whether age, sex and nonblind mortality are predictors of blind deaths.

26.6.1   Summaries of Traditional Regressions

The overall R square value is 0,994, which means 94,4% certainty about the prediction of the outcome by the three predictors. This is significantly different from zero % at p = 0,000. However, on clinical grounds the presence of multicollinearity in the data was suspected. And, therefore, its presence was assessed with one-by-one linear regressions. If T > 0,85, then multicollinearity is in the data, and the model is no longer entirely valid. One of the two variables responsible must be removed. For assessment the underneath commands are required.

The correlation coefficient between gender and ageclass is 1000, much more than 0,85. Also five more one-by-one correlations were over 0,85. The traditional regression model is no longer valid. It model must be replaced with a ridge regression. Ridge regression is a technique for analyzing multiple regression data that suffer from multicollinearity. By adding a degree of bias to the regression estimates, ridge regression reduces the standard errors. It is hoped that the net effect will be to give estimates that are more accurate without loss of sensitivity of testing.

26.6.2   Summaries of Kernel Ridge Regressions

The best fit kernel density model was obtained by the polynomial kernel ridge regression with an R Square value of no less than 1,00 (100% certainty of prediction), which is better fitted than the one from the above flawed traditional linear regression of 0,994. And so, kernel ridge regression provided not only a less flawed analysis, but also provided a better R-square value of prediction of 1,00 instead of 0,994.

We note, that several R squares of the different kernel density models were negative. These values are from ill fitting kernel density models. The phenomenon of negative kernel density R squares is explained in the Chap. 5 entitled "Some terminology".

26.7   Chapter 7

The history, background, and the development of the analytical data models of traditional and kernel regressions have already been addressed in the Chaps. 1, 2 and 3. In this chapter the numbers of stools on a new laxative as outcome and the numbers of stools on the old laxative as predictor in 35 constipated patients will be used as data example. Simple linear regression produced a borderline p-value of 0,049, not a very powerful result. More statistical power was desirable. A GENLIN (Generalized linear regression-generalized linear regression) procedure can be followed using maximum likelihood estimators and/or robust regression.

26.7.1 Summaries of Traditional Regressions

With traditional linear regression the old treatment was a borderline significant predictor at p = 0,048 of the novel treatment. More statistical power was desirable.

Neither was the overall R square value of 0,219 a strong predictor (21,9% certainty about the outcome knowing the predictors. R square values under 25% are considered to be weak predictors. A problem with robust regressions in GENLIN is that no R Square values are provided, and, so, the strength cannot be precisely tested against kernel ridge models. Quantile regression does not provide R Square values, but pseudo R square values can do the job (see the Chap. 5 entitled "Some Terminology). And, so, its datafit can somewhat better be compared against kernel ridge regression and traditional linear regression. The best fit pseudo R square values were with the quantiles 0,1, 0,2, 0,3, and the respective values 0,310, 0.259, and 0,220. All of them were slightly "betterfit" than the traditional regression R square of 0,219. Yet all them were pretty poor predictors giving only 31,1%, 25,9%, and 22,0% certainty about the outcome knowing the predictors, and stronger prediction models are welcome. Kernel ridge regressions will be applied for the purpose next.

26.7.2 Summaries of Kernel Ridge Regressions

Obviously, the polynomial kernel density provided the best fit R square value of no less than 0,831 (83,1% certainty is predicted by the predictors about the outcome. The scatterplot of the predicted newtreat values versus predicted ones shows a very nice linear pattern of the polynomial kernel ridge regression, while the spread of the residuals had a constant pattern.

26.8 Chapter 8

In a 250 patient data file 12 highly expressed genes were tested for effect on a clustered outcome variable of drug efficacy. The multiple variables regression showed that 6 genes were very significant independent predictors of drug efficacy scores. When tested against kernel ridge linear regression, the results of the traditional multiple variables regression and the latter regressions were pretty much similar. Using other kernel density models provided quite better datafit like the Additive Chi2 kernel density, the Laplacian kernel density, and the Polynomial kernel density models.

Better statistics can thus be obtained with the help of kernel ridge regressions using one of the eight kernel density models available in the kernel ridge regression menu of SPSS statistical software.

26.8.1   Summaries of Traditional Regressions

The R square value of the traditional linear regression is 0,729. It means that we have 72,9% certainty about the summaryoutcome knowing the predictors, and we are only 27,1 uncertain. This is a very good predictive result. R squares under 0,25 are very poor, 0,25–50 are reasonable, and over 0,50 are strong predictive models. Yet we may want to assess the predictive potential of the kernel ridge regression models.

26.8.2   Summaries of Kernel Ridge Regressions

The underneath table and graph are in the SPSS output. The R square value of the Linear density model is 0,723, pretty good but not better than the R square value from the traditional linear regression which was 0, 729 (adjusted 0,715).

Better datafits might be obtained by alternative kernel density models. To answer this question, analyses with more kernel density models were performed.

Additive Chi2 kernel density model   R square = 0,766
Chi2 kernel density model   R square = −0,028
Cosine kernel density model   R square = 0,445
Laplacian kernel density model   R square = 0,793
Polynomial kernel density model   R square = 0,951
Radial basis neural network density function   R square = −0,061
Sigmoid kernel density model   R square = 0,000

Three kernel density models provided better fit data statistics with R square values from 0,766 to 0,951. And so, if you need a data result better than that of the linear density kernel ridge model, you may wish to use one of those in your analysis.

26.9   Chapter 9

In a 40 patient study the effect of (1) gender, (2) treatment modalities, and (3) the interaction of the two on the prevention of episodes of paroxysmal atrial fibrillation (paf) was studied. Interaction adjusted multiple variables regression predicted significant effects of gender and interaction on the outcome "prevention of paroxysmal atrial fibrillation at $p = 0,0001$. The predictor treatment modality (metroprolol or verapamil) was not statistically significant.

26.9.1   Summaries of Two Predictor Regressions

The R square was 0,336 meaning that the outcome is predicted by the two predictors gender and treatment with 33,6% certainty, which is a reasonable but not strong predictive power. More power is welcome. Therefore, kernel ridge regressions were performed.

Four kernel density models produced R square values (much) larger than that of the traditional linear regression (0,336).

Chi2   R square 0,654
Laplacian   R square 0,641
Polynomial   R square 0,716
RBF (radial basis function)   R square 0,641.

With the polynomial kernel density model 71,6% certainty about the outcome is given. Obviously the kernel ridge model provided a much better R square value and thus better fit for the data.

26.9.2   Summaries of Three Predictor Regressions

By three predictors 75,8% of certainty about the outcome is given by the traditional linear regression. This is much better than the 33,6% prediction of the two predictor traditional linear regression. The polynomial kernel ridge density model provided the best data fit out of eight alternative kernel density models. An R square of 72.2% is very good and virtually equally powerful as that of the traditional linear model with 75,8%, although not better.

Five kernel density models produced R square values (much) larger than that of the traditional two predictor regression (0,336).

Additive_chi2   R square 0,710
Chi2   R square 0,634
Laplacian   R square 0,680
Radial basis function R square 0,680
Polynomial   R square 0,722.

26.10   Chapter 10

Clinical chemistry has been recognized as helpful as a predictor of morbidity/ mortality scores. In a 200 patient data file of patients with sepsis the effect of laboratory predictors on survival/septic death was assessed. Traditional regression consisted of binary logistic regression with the logodds of survival from sepsis as outcome and the various laboratory values as predictors. Obviously, bilirubine, c-reactive protein, leucos are significant predictors, while the remainder of the predictors were insignificant. This result was pretty disappointing, because clinically we would expect many more laboratory values to predict survival from sepsis. In a kernel ridge regression (Cosine density model) the R square was better sensitive than the Cox and Snell pseudo R square: 0, 811 vs 0,700. The cosine frequency distribution can be observed in the Chap. 5 entitled "Some Terminology".

26.10.1   Summaries of the Traditional Regressions

The test-retest reliability of the manifest variables as assessed with Cronbach's alphas using the correlation coefficients with one variable missing are given. All of the missing data files should produce at least by 80% the same result as those of the non-missing data files (alphas >80%). Indeed, none of the predictor variables after deletion reduced the test-retest reliability. The data are reliable. A binary logistic regression can be performed.

In order to compare the binary outcome logistic model with the kernel ridge model, it would be convenient to compare the R square values of either of the two methods. However, the logistic models do not provide the maths to produce R square values. Instead, however, Cox and Snell (1989) proposed as alternative a pseudo R square based on likelihood statistics. Its upperbound is like that of the true R square 1000. However, It sometimes underestimates the certainty proportion in the binary data as given. In our data example the Cox and Snell pseudo R square equalled 0,700.

26.10.2   Summaries of the Kernel Ridge Regressions

The Cosine kernel density model produced an R square somewhat betterfit than that of the Linear kernel density model, and it was actually the bestfit model as compared with the traditional binary logistic model, 0,811 versus 0, 700.

26.11   Chapter 11

The effect of month on mean c-reactive protein levels was assessed in 18 subsequent months. Usually lag numbers are used for analysis, with, e.g., monthly measures meaning the paired differences between the first month measures as compared to subsequent monthly measures.

26.11.1 Summaries of Traditional Regressions

26.11.1.1 Autocorrelations

The monthly autocorrelation coefficients with their 95% confidence intervals show that the magnitude of the monthly autocorrelations changes sinusoidally. The significant positive autocorrelations at the month no.13 (correlation coefficients of 0,42 (SE 0,14, t-value 3,0, p < 0,01)) further supports seasonality, and so does the pattern of partial autocorrelation coefficients (not shown): it gradually falls, and a partial autocorrelation coefficient of zero is observed one month after month 13. The strength of the seasonality is assessed using the magnitude of $r^2 = 0,42^2 = 0,18$. This would mean that the lag curve predicts the datacurve by only 18%, and, thus, that 82% is unexplained. And so, the seasonality may be statistically significant, but it is pretty weak, and a lot of unexplained variability, otherwise called noise, is in these data.

26.11.1.2 Curvilinear Regressions

The effect of times on C-reactive protein was studied using traditional curvilinear regression. The traditional linear OLS regression provided a very poor fit for the C-reactive protein outcome data. Instead various curvilinear traditional regressions were assessed. A poor fit was shown for all of them. In fact, the best result was obtained by the S model with a p-value, however, of only 0,12 and an R square value of 0,106.

26.11.2 Summaries of Kernel Ridge Regressions

The best fit kernel density model was obtained by he Additive Chi square density model with an R square value of 0,131. This R square value was slightly better than that of the above traditional curvilinear regression S model with an p-value of 0,12 and an R square value of 0,106. This would mean that by the S model 10,6% of the certainty about the outcome was provided, while by the Additive chi square kernel density model 13,1% of the certainty about the outcome was provided.

26.12 Chapter 12

As data example the effect of different dosages of prednisone and beta-agonists on peakflow was measured in 40 patients. The spread of the outcome data may be smaller with low dosage treatments, than it may with high dosage treatment. The effect of prednisone on peak expiratory flow was assumed to be more variable with increasing dosages. Can the spread in the data, therefore, be measured with more precision, if linear regression is replaced with weighted least squares procedure. Does kernel ridge regression provide a meaningful alternative to traditional and weighted least squares regressions.

26.12.1   Summaries of Traditional Regressions

In the traditional linear regression an R square value of 0.582 is obtained, and the linear effects of prednisone dosages are a statistically significant predictor of the peak expiratory flow, but, surprisingly, the beta agonists dosages are not. A weighted least squares analysis was subsequently performed, which is a way for adjusting heteroscedasticity. The output table now shows an R square value of 0.716. It has risen from 0,582, and provides thus better statistical power. The above lower table shows the effects of the two medicine dosages on the peak expiratory flows. The t-values of the medicine predictors have increased from approximately 0,5 and 10 to 3,2 and 14. The p-values correspondingly fell from 0.000 and 0.552 to respectively 0.000 and 0.002. Obviously, after adjustment for heteroscedasticity, the beta agonist became a significant independent determinant of peakflow.

26.12.2   Summaries of Kernel Ridge Regressions

The best fit R square value was obtained from the Laplacian kernel density model, with a magnitude of 0,709, which is almost as large as that of the weighted least square analysis. The advantage of the kernel density analysis is that it is easier and does not require the pretty complex procedure of selection of the weight variable and calculation of likelihoods for different powers. We should add, that one kernel density model, the sigmoid model, produced a negative R square value. Kernel ridge density model can some time be negative. How is it possible that with kernel ridge regression R square values can, obviously, be negative!!! The answer is, that with kernel ridge regression a negative R square value is possible for models where a datafit is worse than horizontal.

26.13   Chapter 13

The effect of race, age, gender on physical strength was assessed in 60 patients. The variable race was assessed as a stepwise variable with four categories (1–4). and also in a slightly restructured model as four binary variables (black yes or no, white yes or no, Asian yes or no, and hispanic yes or no).

26.13.1   Summaries of Traditional Regressions

The R square of the unrestructured model was 0,249, The F statistic and p-value of no difference from zero were 6183 and 0,001. The R square of the restructured model was 0,657. The F statistic and p-value of no difference from zero were 20,728 and 0,000.

26.13.2   Summaries of Kernel Ridge Regressions

The R square of the best fit unrestructured kernel ridge model was 0,393 using the Polynomial kernel density model, which was much better-fitted than that of the unrestructured traditional regression.

The R square of the best fit restructured kernel density model was 0,693, again substantially better than that of the restructured traditional regression.

We should add that many kernel density models produced strong negative R square values as a consequence of poor data fit of these models. The somewhat awkward result of negative R square values is observed with kernel ridge models where the data fit is worse than horizontal.

26.14   Chapter 14

In a 20 patients parallel group trial the effect of treatment modality (sleeping pill or placebo) and other covariates on hours of sleep was studied.

26.14.1   Summaries of Traditional Linear Regressions

The overall R square value was 0,970, meaning that the outcome is predicted by the predictors by 97%. This 97% predictive certainty was significantly different from a 0% predictive certainty at $p < 0.000$. However, on clinical grounds the presence of multicollinearity was suspected. And, therefore, its presence was assessed with one-by-one linear regressions: if $R > 0,85$, then its presence is confirmed, and the model is no longer valid. One of the two variables responsible must be removed.

26.14.2 Summaries of Kernel Ridge Regressions

The R square of the Linear kernel density model was 0,963 which demonstrates a very good data fit virtually similar to that of the traditional multiple variables regression of 0,970. But kernel ridge regression is better, because it is adjusted for multicollinearity. Also in the output is a graph of a scatterplot of predicted values versus treatment effects. A very nice linear data pattern is observed in the underneath graph.

26.15 Chapter 15

In a 20 patients parallel group trial the effect of treatment modality (sleeping pill or placebo) and other covariates on hours of sleep was studied.

26.15.1 Summaries of Traditional Linear Regressions

The overall R square value was 0,970, meaning that the outcome is predicted by the predictors by 97%. This 97% predictive certainty was significantly different from a 0% predictive certainty at $p < 0.000$. However, on clinical grounds the presence of multicollinearity was suspected. And, therefore, its presence was assessed with one-by-one linear regressions: if $R > 0,85$ (or $< -0,85$), then its presence will be confirmed, and the model will no longer be valid. One of the two variables responsible must be removed.

26.15.2 Summaries of Kernel Ridge Regressions

The R square of the Linear kernel density model was 0,963 which demonstrates a very good data fit virtually similar to that of the traditional multiple variables regression of 0,970. But kernel ridge regression is better, because it is adjusted for multicollinearity. Also in the output is a graph of a scatterplot of predicted values versus treatment effects. A very nice linear data pattern is observed in the underneath graph.

26.16   Chapter 16

With traditional multivariables regression the outcome should be continuous, and in the example of the current Chap. the outcome qol score is measured in only 5 steps. Nonetheless the SPSS program provides a result of such an analysis. However, because of model's poor fit the R square is small, only 0,105. Better datafit should be obtained by several alternative analysis assessments. As data example the effect of three predictors on health related quality of life was assessed with traditional and kernel ridge regressions.

26.16.1   Summaries of Traditional Regressions

The R square value of the traditional linear regression was 0,105, which indicates 10,5% certainty to predict the outcome qol score. This is a poor result. R square values under 0,25 are assumed to have very poor predictive potential. The usually applied Cox and Snell R Square of the multinomial regression was larger than the R square computed by the traditional linear regression (0,191 vs 0,105). The overall Cox and Snell pseudo R square of the ordinal regression model was only 0,117, smaller than that of the multinomial regression (0,191).

26.16.2   Summaries of Kernel Density Regressions

Kernel ridge density models show, that, particularly.

the Chi2 kernel density model (R square 0,241),
the Laplacian kernel density model (R square 0,262),
the Polynomial kernel density model (R square 0,205),
the RBF kernel density model (R square 0,237)

performed (much) better than did the above traditional regression methodologies. A kernel density R square over 0,250 like the one with the Laplacian model is assumed to be a reasonable result in terms of predictive performance.

26.17   Chapter 17

Outcome categories can be assessed with multinomial logistic regression. With categories both in the outcome and as predictors, the latter models assume, that for each predictor category or combination of categories x_1, x_2,... slightly different a-values are computed with a better fit for the outcome category y than a single a-value $y = a + b_1 x_1 + b_2 x_2 + \ldots$.

We should add that, instead of the above linear equation, even better results may be obtained with log-transformed outcome variables (log = natural logarithm).

$$\log\ y = a + b_1 x_1 + b_2 x_2 + \ldots.$$

As a 55 patient study example, three hospital departments (no surgery, little surgery, lot of surgery), and three patient age classes (young, middle, old) were the predictors of the risk class of falling out of bed (fall out of bed:

"no",
"yes but no injury",
"yes and injury").

Kernel ridge regression can also handle binary or multinomial outcome data. And despite the dicho- or multitomous outcomes, no pseudo R squares are needed, but, rather, real kernel ridge R square values can be obtained.

26.17.1   Summary of Traditional Regression with Multinomial Logistic Regression

The multinomial model shows, that the department is a very significant predictor of fallingoutofbed, and also ageclass is a significant factor with p-values <0.000 compatible with the very high pseudo R squares of 0,718 and 0,594.

26.17.2   Summary of Kernel Ridge Regression

The best fit kernel density model was here the Polynomial with R square 0,771. a linear model in scatterplot of "falloutofbed-data" by "predicted data" was observed in the kernel ridge Polynomial kernel density model.

More well fitting kernel density R square values were the

Additive_chi2 kernel ridge density model with R square 0,748,
Chi2 kernel ridge density model with R square 0,756,
RBF kernel ridge density model with R square 0, 756.

26.18   Chapter 18

With decision trees, data samples of patients with and without the presence of a disease are assessed for subgroup properties. Usually binary variables are used for assessment, but binary cut-off values of continuous variables can also be used. Linear regression can be applied for finding the optimal cut-offs of subgroups, i.e., the cut-offs with the linear regression equation, that produces the largest test statistic. It is a lot of work, and it is, sometimes, called exhaustive searching, but for a computer it is not hard to do. We should add, that the computer is even capable of finding best cut-offs with partitioning into more than two subgroups, if required. But we will stick to two subgroups, in order to avoid too much complexity in the models. Instead of or in addition to regression-tree-analysis quantile regressions may be helpful to provide additional significances of the predictor variables unobserved in the ordinary least squares regression and/or in the entire tree regression model. A 953 patient data file is used of various predictors of LDL (low density lipoprotein) - cholesterol reduction including weight reduction, gender, sport, treatment level, diet. The weight reduction and sport were very significant independent predictors of LDL cholesterol reduction.

26.18.1   Summaries of the Traditional Multiple Variables Linear Regressions

The traditional R-square is 0,972, a very good predictive model corresponding with an ANOVA (analysis of variance) with an F test of 662,847 and a p-value <0,000. In the coefficients table of the traditional ordinary least regression analysis, particularly weight reduction and sport activities were very significant predictors, three more independent variable were insignificant. Probably, interactions between the five independent variable were responsible. Two methods are relevant for further assessment. In the regression tree only weight reduction significantly contributed to the model (p = 0,094), with the overall mean and standard deviation dependent variable ldl-cholesterol.

26.18.2   Summaries of Kernel Ridge Regressions

The best fit predictive model was given by the Laplacian kernel density model with an R square value of 0,981. A very poor predictive model was the sigmoid kernel density model with a very large negative R square value (see the Chap. 5 "Some Terminology" for explanation of negative R square values).terol in the parent (root) node. And so, decision tree analysis contributed little significance to the overall analysis. The scatterplot of the Laplacian kernel ridge density model shows a very close linear pattern. A nice linear scatterplot gives the predicted values vs the predicted residuals. When returning to the dataview screen, a 7th data column giving the predicted ldl_reductions by the Laplacian kernel ridge density model is observed.

26.19 Chapter 19

In the current chapter the effect on measured body surface of predictors including gender, age, weight, and height were assessed in 90 patients.

26.19.1 Summaries of the Traditional Multiple Variables Linear Regression

The traditional linear regression with four predictors produced an R square of no less than 0,996. Thus, the predictors together explained the outcome by 99,6%. A very good result. The insignificant gender effect and the weakly significant age effect might have been caused by multicollinearities or interactions. As kernel ridge regression adjusts for multicollinearity, we will now assess kernel ridge regressions of the same data.

26.19.2 Summaries of the Kernel Ridge Regression

A negative R square like observed above is sometimes observed with kernel ridge density modeling. See also the Chap. 5 "Some Terminology". In the underneath scatterplot the predicted values using the best fit kernel density model, the Polynomial kernel density model (R square 0,997). More very strong predicting kernel density models were observed, the Linear, the Additive_chi2, and the Chi2 kernel density models. A very good linear data pattern is observed above with a Polynomial kernel density model. Underneath it is also seen that residuals have a constant pattern as it should, and do not increase with increasing predicted values. When returning to the dataview screen it is observed that SPSS has provided predicted values for all of the patients.

26.20 Chapter 20

In 139 physicians the effect on giving lifestyle advise or not of various predictors is assessed.

26.20.1   Summaries of Traditional Multiple Variables Logistic Regressions

The variable "physician" is the physician identity number. Each physician produced two rows of data, one with and one without prior education as outcome. The predictor variables were physicians' identity, age, prior postgraduate education or not, and countrypractice or not. Only the physician id and having had prior postgraduate education were significant predictors.

The R square is not easily obtained from binary outcome data like logodds, but, instead, SPSS provides pseudo R squares obtained from loglikelihood statistics. The Cox and Snell R square value was 0,579, which is a pretty good value for making reliable predictions. It means that 57,9% certainty for predicting the outcome should be given by the statistical model given. The Nagelkerke pseudo R square is even better (0,785) although infrequently used.

26.20.2   Summaries of Kernel Ridge Regressions

The effect of physician education on lifestyle advise is visualized by a series of strong positive R square values in various kernel density models. The best fit model is the Chi2 kernel density model, with R square 0,865. The model provides 86.5% certainty by the predictors about the outcome. However, more kernel density models were strong predictors too. The Chi2 kernel density model R square of 0,865 (86,5% certainty) is much better than the pseudo R Square value of the Cox and Snell approach of only 0,579, i.e., 57,9%.

More powerful kernel ridge regression R square predictors were:

Linear kernel ridge regression   R square 0,598
Additive_chi2 kernel ridge regression   R square 0,664
Laplacian kernel ridge regression   R square 0,750
Polynomial kernel ridge regression   R square 0,709
RBF kernel ridge ridge regression   R square 0, 656.

All of them were better sensitive than the pseudo R square value of Cox and Snell of 0,579 as established in the traditional binary logistic regression analysis of the same data.

26.21   Chapter 21

This chapter addressed the performance of kernel ridge regression versus traditional linear regression and a form of linear regression with time as weighting factor will be tested. As an example 50 patients were studied for numbers of paroxysmal atrial fibrillations (PAFs), while on a parallel-group treatment with two different treatments. The scientific questions were: do psychological and social factor scores and different treatments affect the numbers of PAFs per person per period of time. A weighted least squares (WLS) regression provided a borderline significant effect of treatment modality on number of PAFs per person per period of time.

26.21.1   Summary of Traditional Regressions

Treatment modality is weakly significant, and psychological and social score are not. Furthermore, days of observation is very significant. However, it is not very precise to include this variable if your outcome is the numbers of events per person per time unit. Therefore, we will adjust the outcome variable for the differences in days of observation using weighted least square (WLS) regression.

The traditional linear regression provided an R square value of 0,603, pretty good, and meaning that the outcome can be predicted by the predictors by over 60%. The weighted least square method adjusts for the days of observation and is a more precise analysis for our purpose. Unfortunately, the adjustment caused the predictive property of the model to fall from R square = 0,603 to only R square = 0,091. Certainty of prediction is only left by 9,1%, a very poor result in this analysis.

26.21.2   Summary of Kernel Ridge Regressions

In order to assess, whether kernel ridge regression provided better fit data modeling different kernel density models were performed. Note, that SPSS version 28.0.1.0 is required, because earlier version of SPSS statistical software do not carry kernel ridge regressions. The best fit kernel density models were.

Additive_chi2 kernel density model	R square 0,834
Chi2 kernel density model	R square 0,657
Laplacian kernel density model	R square 0,580
Polynomial kernel density model	R square 0,975
RBF kernel density model	R square 0,564.

Of the above 5 kernel density models 3 were, thus, better sensitive predictors than were the traditional linear regression with an R square value of 0,603 and weighted least square regression with an R square of 0,091.

26.22   Chapter 22

A data example of 3390 patients with epilepsy was studied for various predictors of convulsions. The data file was partially obtained from the "SPSS Samples" site.

26.22.1   Summaries of Traditional Multiple Variables Linear Regressions

First a data model with three predictors was analyzed with traditional multiple variables linear regression, but the predictive potential was pretty poor with an overall R square of only 0,003, and a p-value of p = 0,015. Second, a four predictor model was assessed. It performed slightly better with an R square of 0,007, and a p-value of <0,001.

26.22.2   Summaries of Kernel Ridge Linear Regressions

The kernel ridge regression of the four predictor model produced a much better R square value of 0,806 (80,6% predictive certainty) in the Laplacian kernel density model. A very good linear data fit of the predicted values versus the measured values (numbers of convulsions) was observed.

26.23   Chapter 23

A data example of 252 persons was studied for the effects of different foods on breakfast scenario taken.

26.23.1   Summaries of the Traditional Multiple Variables Linear Regression and Multinomial Logistic Regression

The R square value of 0,122 can be interpreted as a 12,2% certainty about the outcome given by all of the predictors simultaneously. This is less than 25%, and it means, that the model is a poor predictive model. The analysis of variance (ANOVA) table produced a F (Fisher) statistic of 4236 compatible with a p-value of <0,001. As an alternative multinomial logistic regression may provide better fit results. With multinomial logistic regression, indeed various predictor variables were significant predictors of different outcome values, the menu scenarios. However, for comparisons with kernel ridge regression we are particularly interested in the overall R square values, giving the percentage of certainty of the predictors on the outcome. However, traditional R square values are not provided by the multinomial model. Fortunately, pseudo R square based on log likelihood statistics are provided instead. Cox and Snell pseudo R squares are commonly used. The value of 0,400 is a lot better fit than the R square = 0,112 of the traditional linear regression of the data.

26.23.2   Summaries of the Kernel Ridge Regressions

The best data fits were provided by the Laplacian and the Polynomial kernel density models (respectively R square values 0,510 and 0,659, meaning 51,0% and 65,9% predictive certainties). The traditional multiple variables linear regression and multinomial model provided R square values of only 0,122 (12,2% predictive certainty), and 0,400 (40% predictive certainty.

Note, that the RBF (radial basis function) kernel density model produced a negative R square. Explanation is given in the Chap. 5, "Some Terminology".

26.24   Chapter 24

The effect of personal factors on anorexia was studied in 217 persons.

26.24.1   Summaries of the Traditional Multiple Variables Linear Regression

The R square value of 0,481 can be interpreted as a 48,1% certainty about the outcome given by all of the personal factors simultaneously. This is close to 50%, and it means that the model is a reasonable, but not yet a strong prediction model. The analysis of variance (ANOVA) table produced a F (Fisher) statistic of 17,278 compatible with a p-value of <0,001. The coefficients table showed which of the predictor variables were the strongest predictors. Six of the 11 predictors were statistically significant at a $p < 0,10$ level.

26.24.2   Summaries of the Kernel Ridge Regression

The best fit model was provided by the Polynomial kernel density model (R square 0,745), and this was much better than the traditional linear regression R square value of 0,481.

More very good kernel density models were observed. For example:

the RBF kernel density model: R square 0,625
the Laplacian kernel density model: R square 0,620
the Chi2 kernel density model: R square 0,663
the Additive_chi2 kernel density model: R square 0,509.

All of them were better fit than the traditional linear regression with an R square value of 0,481.

26.25   Chapter 25

In this chapter the effect on weight loss of physical exercise, calorieintake, and age was studied in 64 patients. The outcome of clinical research is, generally, affected by many more factors than a single one, and multiple variables regression assumes, that these factors act independently of one another, but why should they not affect one another. With multicollinearity one or more x-variables have a strong correlation with one another. How does ridge regression deal with multicollinearity? Ridge Regression is a technique for analyzing multiple regression data that suffer from multicollinearity. **By adding a degree of bias to the regression estimates**, ridge regression reduces the standard errors. It is then hoped, that the net effect will be more adequate.

26.25.1   Summaries of the Traditional Multiple Variables Linear Regression

The R square value of 0,502 can be interpreted as a 50,2% certainty about the outcome given by all of the predictors simultaneously. This is close to 50%, and it means that the model is a reasonable, though not a strong prediction model. The analysis of variance (ANOVA) table produced an F (Fisher) statistic of 14,890 compatible with a p-value of <0,000. Obviously, weight loss is significantly dependent on the combination of all of the x-variables.

The coefficients table shows which of the predictor variables are the strongest predictors. Calorieintake and interaction between physical exercise and interaction between physical exercise and calorieintake are the strongest predictors and significantly different from zero correlations at p-values of <0,000 and < 0,002. In order to find out whether kernel ridge regression is able to provide a still better predictive score for this data, we will, subsequently, perform kernel ridge regressions.

26.25.2   Summaries of Kernel Ridge Regressions

The R square values of some kernel density models were considerably better than the R square of the traditional linear regression (0,502). For example, 0,641 for the polynomial kernel density model and 0, 677 for the Additive_chi2 kernel density model, and so kernel density provided better data fit models than did traditional linear regression.

Printed in the United States
by Baker & Taylor Publisher Services